Bees: Biology & Management

By

PETER G. KEVAN

Cover design by Andrew Morse with photographs and drawings from the Anthecology Laboratory at the University of Guelph, Ontario, Canada.

Reprinted with minor corrections & additions, June 2007

ISBN 0-9680123-4-5

Enviroquest Ltd., 352 River Road, Cambridge, Ontario N3C 2B7 Canada

Dedication & Acknowledgements

Dedication

I dedicate this book to my fond memories of John Stephen King'ang'i Mbaya, simultaneously for each of us, Father and Son, and my graduate student.

Mbaya, J. S. K. 1993.   Resource hoarding and defensive behaviours of African honey bees (*Apis mellifera* L.) in Kenya. Ph. D. Dissertation, University of Guelph.

Acknowledgements

I thank all the students who, over many years, have provided such enthusiastic interest in bees, honeybees, and basic insect biology. I thank my numerous friends and colleagues who have contributed so much to my understanding and appreciation of the marvels of bees, pollination, and general entomology. I especially appreciate the guidance and instruction I have received from so many beekeepers over the years. The Ontario Beekeepers' Association has been especially encouraging, but before that, the Pikes Peak Beekeepers' Association provided me with memorable opportunities to share my knowledge and enthusiasm. My students have all been supportive, and especially I thank Svenja Belaoussoff and Mark Robinson for help in preparing this book. However, before this book was possible, Eva Miciula Dabrawska and Arthur Miciula provided invaluable contributions through assembling information, organising notes, and making the illustrations. I thank them sincerely. Lenore Latta has assisted greatly through her keen eye for detail in technical editing. Andrew Morse kindly applied his artistic side to the cover design. Recently, with the help of Dr. Sigrun (Ziggy) Kullik, the teaching team from Winter, 2007 and the students, minor corrections and additions have been made. The University of Guelph provided some financial support towards preparation of this book in conjunction with the course "Introductory Apiculture".

Credits for Illustrations and Tabulated Data

Many people and organizations have allowed the use of their illustrations, adapted, redrawn or in the original. I thank them all, and if any are not properly credited, please inform me so that corrections can be made. A listing of the credits is given at the end of the book.

TABLE OF CONTENTS

PREFACE

The University of Guelph has a long tradition in apicultural education. This book was developed from lecture notes and other material over the last two decades, but the course has been taught annually for over a century and a quarter. Rev. F. W. Clarke gave the first lectures in 1893.

Although the course on which this book is based is called "*Introductory Apiculture*," a better name would be "*Bee Biology: Principles & Applications*" or "*Bees: Biology and Management.*" Over the years, and as the Ontario Agricultural College has changed to meet increasingly rigorous and academic standards, the course's flavour has come to embrace more basic bee biology and to relate that to applications in apiculture. Apiculture itself has changed greatly over the past 170 years, and the course reflects that too. This book, and the course, is not about "How to keep bees" but why bees are kept, what it is about them that makes them amenable to management and domestication: their anatomy, physiology, behaviour, diversity, ecology, and value to agriculture and the natural world. Of course, the main study organisms are European races of the Western honeybee, *Apis mellifera*, but other bees are far from forgotten.

The course content, and order of presentation of information in the book, runs as follows. The first part of the course is introductory to castes and anatomy of bees (SECTIONS I to III). That provides a framework by which to understand colony structure and bees as micromanipulators of their own world. Thus, greater discussion can be developed about bees' roles in the colony, division of labour, and activity changes with age (SECTIONS IV to VI). The concepts of insect metamorphosis and development are described (Chapter IV – 12). Not all insects pass through the four stages of egg, larvae, pupa, and imago (adult) as do bees. Moreover, honeybees develop into their different castes according to genetic programming; the drones have only half the diploid set of chromosomes of the female queen and workers (Chapter VI – 22); and nutrition is provided by the worker bees (Chapters IV – 12 & 13). Development is under hormonal control, as is age-dependent behaviour (Chapters IV – 14 & 15). The colony itself proceeds through a seasonal cycle, and in this book the example of the northern temperate zone is used (Chapters V – 20, VI – 21, & VIII – 32). The interesting features of bees' mating strategies and genetics (Chapters VI – 21 to 23) combine naturally with colonial reproduction, swarming, supersedure, absconding, laying-workers, and migration (Chapter V – 20), and with those activities associated with food-gathering behaviour and efficiency through optimal foraging theory (Chapter V – 18).

No course in bee biology is complete without a thorough description of the remarkable ways that bees communicate (Chapters V – 16 & 17) by chemistry (pheromones) and by dance about conditions within the hive and outside, including their abilities in navigation (Chapters V – 17 & 18). Scientific controversies about these aspects of bee biology illustrate the value of hypothetico-deductive reasoning and rigorous evaluation of ideas in science. Remarkable is the capacity for honeybees, and other bees, to incubate their brood and stay warm over winter (Chapter V – 19).

Insects' nervous systems (Chapter III – 9) show many functional similarities to our own, in spite of being rather different in structure. The brain, made up of about 850,000 cells, and segmental ganglia are involved not only in neural processing, but also in chemical control of development and behaviour through hormonal regulation (Chapter IV – 14). Sensory systems and organs, such as vision, olfaction, taste, mechanical senses, and others (SECTION VII) are linked through the brain, with inputs that are stored in memory and used in complex behaviours, including an amazing capacity for learning.

The order in which the chapters are presented represents only one way in which honeybee biology, ecology and behaviour can be integrated. Throughout the book, there are many cross-references made to allow readers to follow their own leads to understanding the complex whole.

From a strong biological basis, the book continues into applications in beekeeping. The historical roots are followed from the Stone Age to modern apiculture (Chapter VIII – 31). The year as a beekeeper may experience it is described briefly (Chapter VIII – 32), relying heavily on basic information presented in SECTIONS IV to V and VI, in part. As with all domesticated organisms, bees are not without afflictions of diseases, predators and pests (SECTION IX).

The final topics are about hive products, the nature of honey, bees' wax, pollen, royal jelly, and the like (SECTION X), and the far more valuable service that bees provide to agriculture through pollination (Chapter X – 45). The value of bees and hive products is huge on a global scale, and Canada ranks high in beekeeping expertise and innovation as well as in apicultural production. Pollination services to agriculture, and, in some parts of the world, to nature, rely heavily on honeybees. However, other valued contributors to global productivity, food security, and environmental sustainability are other pollinators (SECTION XI). Some have been integrated in agriculture, and the conservation of all pollinators is now a matter of international concern (Chapter XI – 51).

To complement this book, a series of on-line lectures is planned. Those are expected to further integrate information. Library resources should be consulted in accordance with the SECTIONS and Chapters, so a brief bibliography of widely available publications is provided after Chapter XI – 51.

Most beekeepers are friendly and only too pleased to have interested people visit them in their bee yards and honey houses. Please try to visit a beekeeper and see for yourself how honey reaches your dining room table.

Peter G. Kevan, Professor, University of Guelph
18 December 1997, revised 20 November 2006

SECTION I

Bees from the Outside:
General External Anatomy, Body Segment by Body Segment
Head, Thorax, & Abdomen

This first section explores how an insect, and particularly a honeybee, is "put together" from an outward viewpoint. The main body segments, the **head**, **thorax**, and **abdomen,** are investigated as to form and function.

The **head**, with its conspicuous appendages for feeding, the **mouthparts**, and its complex sensor array of five **eyes** of two kinds (two large **compound eyes** and three smaller **ocelli**), the **antennae** (or feelers), and organs of **smell**, **taste**, **touch**, **hearing**, and more make the head the sensory and feeding centre. Those organs and their parts in bees are similar to those found throughout the insects. More details on how they work are presented in SECTION VII and in Chapter II – 5 on the processing of food.

The **thorax**, actually made up of three segments itself, contains the engine for locomotion. It has five pairs of appendages, three pairs of **legs** and two pairs of **wings**. That is the base plan for all winged insects. The **legs** are for walking, gripping, and landing. Each leg is made up of many segments, and the **toes** are fascinating. With three legs in contact with the ground at all times, a walking insect is hard to push over. Bees have specializations on their fore and hind legs for harvesting pollen from flowers. Their **wings**, really extensions of their body wall, on each side hook together when they are extended for flight, so the fore and hind wing together act as a single wing. How insects fly is a complex matter. The machinery that the muscular engine operates is an amazing system of mechanical joints, springs, and boxes. The section in Chapter 3 on flight is quite difficult and may require extra study; for those students who become intrigued enough to study more about flight, beware, because Chapter 3 has been simplified.

The **abdomen** comprises yet more segments. The first is specialized, and appears to make up part of the thorax. Most of the **six visible segments** (actually II to VII) look similar. Then tucked away inside the tip of the abdomen are the highly reduced segments that make up the **sting** apparatus in female bees, and the male **intromittent** (meaning "put between") **organ**, or **endophallus**. How the stinger works is described as even more complex machinery of plates, hinges, springs, and needles. How the explosive intromittent organ of the drones works is another story (Chapter VI – 21: "Mating").

Section I – Bees from the Outside: General External Anatomy, Body Segment by Body Segment
Chapter I
Introducing Bees: Their Place in the World of Life, General Anatomy, Exoskeleton & Some Basic Concepts

CHAPTER I – 1

INTRODUCING BEES:
THEIR PLACE IN THE WORLD OF LIFE, GENERAL ANATOMY, EXOSKELETON, & SOME BASIC CONCEPTS

What is it about bees that make them so interesting? To most people, hive products, especially honey, spring to mind. To others, stings are first. Human interactions with these fascinating creatures, especially honeybees, are ancient. Even though there are several species of honeybees (Chapter VI – 25), and within the Western honeybee that is familiar in around the world there are many races (including the infamous "killer" or Africanized honeybee) (Chapter VI – 24), the world of bees comprises as many as 20 to 40,000 species. So, what is a species and what is a bee?

Bees can be thought of as specialized vegetarian wasps. They eat pollen, not meat, for protein (Chapter VI –25). They get protein from flowers (Chapter X –40) where they also obtain nectar, their energy fuel for flying (Chapters I – 3, V –18 & VI – 20) and other activities (Chapters III – 10, IV – 13, V – 17 to 20). They are the micro-manipulators of flowers and bring about pollination, a process vital to human food and fibre production and to the sustainability of natural land-based ecosystems (Chapter X – 45). Some bees can be kept (Chapters XI – 46 to 51), but most are truly wild animals.

This book is mostly about the bees that human beings use. Bees are insects, they are animals, and they are part of the web of life on planet Earth. So how do they fit into the diversity of life? The science of **taxonomy** is the classification of life. It also includes the defining, describing and naming of species. At the level of species, organisms are referred to by two names, i.e., *Genus* and *species* of the **binomial system,** which together make up the scientific name. Simply put, species are reproductively isolated from each other: most can not interbreed at all, or at least if they do, fail to produce viable offspring. Chapters VI – 24 & 25 discuss the taxonomy of bees from the level of **Order** through *genus* and *species* to *subspecies* and **races** (see also below).

Kingdom (now under major rearrangement as a result of modern molecular biological findings)
Phylum (the Arthropoda: from spiders, mites, to crustaceans and insects)
Class (the Insecta)
Order (the ants, bees, wasps and others: Hymeoptera)
Family (there are several Families of bees (Chapter XX))
Genus
Species
Subspecies or Race
(Variety or Strain)

Section I – Bees from the Outside: General External Anatomy, Body Segment by Body Segment
Chapter 1
Introducing Bees: Their Place in the World of Life, General Anatomy, Exoskeleton & Some Basic Concepts

The molecular basis of life resides within the **cells** of living organisms. In summary, the simplest cells are prokaryotic. They lack a nucleus and membrane-bound organelles. Bacteria fall into this group of life forms. Eukaryotic cells, or "true cells," have a **nucleus** and membrane-bound organelles. They are the building blocks of complex life forms, such as plants, animals, and fungi. Animals, such as bees and human beings, are eukaryotic, heterotrophic, motile, and multicellular, with complex tissues, organs and systems. As animals are seen to be more complex, they are considered to be more advanced. It is reasonable to ask: "Are insects, particularly bees, more advanced than human beings?"

The genetic material of all living organisms, both prokaryotes and eukaryotes, is **DNA** (Chapter VI – 23). Virus particles do not have DNA, but do contain RNA as their genetic material. Virus particles are not considered to be really living but some are important as pathogens of bees (Chapters IX – 33 & 34).

DNA is an abbreviation for the molecule deoxyribonucleic acid. It is a major component of **chromosomes.** The elucidation of its chemical and physical structure, the double helix, has changed how human beings view and use the world and the life upon it. The DNA of bacteria is located in a central region of the cell, called the **nucleoid,** where it forms a compact double-stranded loop. The **nuclear chromosomes** of eukaryotes are complexes of DNA and chromosomal proteins (largely lacking bacterial nucleoids). Each chromosome consists of one huge linear, double-stranded DNA molecule running throughout its length. A cell's DNA contains all the genetic material necessary for it to reproduce itself and control its life activities. That, in turn, controls the growth, development, form, and activity of the multicellular organism. The relevance of some of these points is explained in Chapter VI – 23 on genetics, and in Chapters IX – 34 & 35 on bee diseases. Recent research has elucidated the nature of the entire **genome** (the total DNA complement) of human beings, and of honeybees.

The phylum of concern to this book is the **Arthropoda**, but another of indirect concern is the Chordata to which we, as human beings along with other mammals, birds, reptiles, amphibian, and fish, belong. The **Arthropoda** comprises most of the animal species, exhibiting such a marvelous variety of life styles that they are adapted to almost every environment.

The body plan that evolved in arthropods provides for great adaptability, and hence evolutionary success. Arthropods have a principal body cavity, a *true* **coelom** (as do human beings) (see Chapter IV – 12), and **are segmented**. The specialization of appendages, development of skeletal muscle, which allows more rapid response to stimuli than does smooth muscle, and a well-developed nervous system contribute to the success of this Phylum.

Section I – Bees from the Outside: General External Anatomy, Body Segment by Body Segment
Chapter I
Introducing Bees: Their Place in the World of Life, General Anatomy, Exoskeleton & Some Basic Concepts

Arthropods are further characterized by having jointed (as the prefix "arthro" [as in arthritis, inflammation of the joints] indicates) legs (as the suffix "poda" indicates) and an **exoskeleton** (i.e., the skeleton on the outside of the body, similar to a suit of armor).

Having an **exoskeleton** has its disadvantages. Arthropods (consisting of insects, Crustacea [e.g. shrimps, lobster, crabs], Arachnida [e.g. spiders, scorpions], etc.) are, in essence, encased in suits of armor and cannot grow continually. To get bigger, they must shed their exoskeleton, and grow a newer, larger one. Other disadvantages of an exoskeleton are that it restricts how to get information from the outside world within, (i.e., how to see, touch and smell, for examples).

The advantage of a suit of armor lies in its protective value, including against loss of water. Of course, the armor cannot be entirely inflexible. The exoskeleton must have joints by which the body parts can move against each other and must allow for coordination. Just as bones provide support and attachments for muscles in vertebrates, so does the exoskeleton in arthropods. For insects, with all their remarkable abilities to move by walking, running, and flying, and coordinating all their activities from feeding to sex, the exoskeleton is a marvel of natural structural engineering.

The **exoskeleton** of arthropods is made up of a **chitin** base. In Crustacea, minerals incorporated into the exoskeleton make for a hard and crusty shell, but in insects this is not the case. **Chitin** is a polymer of **N-acetyl-glucosamine** (Figure 1-1).

In general, chitin is tough. Some chitin is hard, like that in a beetle's back and that making up the segmented limbs of insects. Some chitin is flexible, as in the wings of insects, and some is both flexible and elastic, allowing movements between the body segments and at the joints.

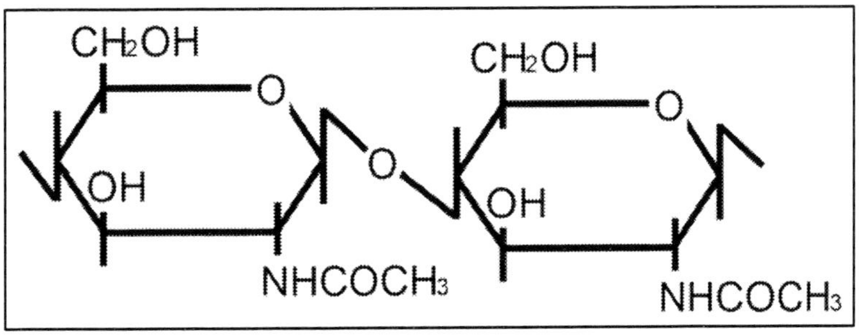

Figure 1-1. Generalized chemical structure of chitin: N-acetyl-glucosamine. The N-acetyl part of the name of this compound means that an acetyl group, $-COCH_3$, is attached to the main structure (glucose) through a nitrogen (N) atom.

After arthropods **moult** to grow (i.e., shed the old exoskeleton to expand into a new one), at first the chitin is soft and elastic. With time, in insects, it darkens and hardens through contact with atmospheric oxygen. This is called **tanning** or **sclerotization**. Chapter IV – 14 presents how moulting is coordinated and the stages by

Section I – Bees from the Outside: General External Anatomy, Body Segment by Body Segment
Chapter 1
Introducing Bees: Their Place in the World of Life, General Anatomy, Exoskeleton & Some Basic Concepts

which it proceeds so that wastage and risks are minimized. Young adult bees are pale and quite soft. Their stings are too flexible to be functional weapons. Such bees are called **callows**.

Getting to Bees and What Makes a Bee a Bee

Within the Arthropoda is the Class **Insecta**. Insects are characterized by having three major body segments (Figure 1-2).

I. Head (the nervous and sensory centre)
II. Thorax (the locomotory centre)
   Appendages: 3 pairs of legs and 2 pairs of wings
   (note that legs and wings are all locomotory appendages:
   legs and wings are not homologous [i.e., made up of the same sets
   of parts] appendages).
III. Abdomen (the digestive, reproductive, and circulatory centres)
   Internal organs mostly in abdomen

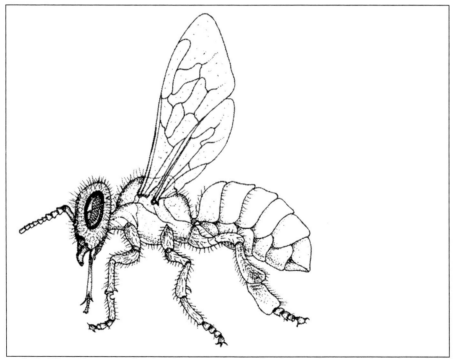

Figure 1-2. The three major body segments of a honeybee worker: head (Chapter I – 2), thorax (Chapter I – 3) and abdomen (Chapter I – 4).

Within the Class Insecta are over 30 Orders. Most of the names of the insect orders end in the suffix "ptera," which derives from the Greek for "wings" (as in "helico*pter*"). Thus, for example, the familiar grasshoppers are in the Order Orthoptera (straight winged), true flies in the Order Diptera (two winged), and butterflies and moths are in the Order Lepidoptera (scaly winged). The ants, bees and wasps, as well as the

Section I – Bees from the Outside: General External Anatomy, Body Segment by Body Segment
Chapter I
Introducing Bees: Their Place in the World of Life, General Anatomy, Exoskeleton & Some Basic Concepts

sawflies and their allies, are placed in the Hymenoptera (membrane-winged insects).

In the **Hymenoptera** are insects characterized by having two pairs of membranous wings and an ovipositor modified into a sting or drill. Included are sawflies, parasitic wasps, other wasps, ants, and bees. One may ask about the ants, as mostly they lack wings. However, during the sexual activity, both the males and females (incipient queens) take to the air on membranous wings, often in huge swarms. The so-called Aculeata (a section of the Hymenoptera) includes the ants, bees and wasps, all with ovipositors (egg-laying organs [see Chapters I – 4 and III – 11]) that are modified into stings, even if the stings are non-functional and vestigial in some groups.

As the science of bee taxonomy has progressed, it has become increasingly difficult to define exactly what separates bees from wasps. In fact, some predatory wasps are so similar to bees that they are included in the hymenopteran Superfamily Apoidea, or, on the other hand, bees are included in the wasp Superfamily Sphecoidea. Within the Superfamily are two groups, the Spheciformes and the Apiformes. Bees belong to the latter group.

Thus, **bees** belong to the hymenopteran **Superfamily Apoidea** and within that the **Apiformes.** All members of the Apiformes are characterized by having branched body hairs. They are thought to be an adaptation for gathering and carrying pollen. Most have a hind basitarsus (part of the hind leg) that is broader than the lower tarsal segments of the leg, and a sting (even if vestigial and non-functional as in the so-called "stingless bees" [Chapter VI – 25]). Bees' mouthparts are mostly longer than those in spheciform wasps. A major difference between the bees and the wasps is that bees are almost entirely herbivorous, deriving their nutrition from flowers as pollen and nectar (Chapter IV – 13). Wasps are mostly carnivores. Within the Apiformes are several **Families** which include honeybees, bumblebees, stingless bees, orchid bees, sweat bees, mason bees, leafcutter bees, orchard bees, and squash bees (Chapters VI – 25 & XI – 46 to 51).

Honeybees, bumblebees, stingless bees, and others belong to the Family **Apidae.** The Family can be broken down to **Tribes,** such as the **Apini (honeybees), Bombini (bumblebees)**, and **Meliponini (the stingless bees).** Those three Tribes are pointed out because they are **eusocial**; that is to say, members live in colonies with three castes (Figure 1-3): the **queen** (fertile female), the **workers** (sterile females), and the **drones** (fertile males) (see Chapter VI – 22). Eusociality is not restricted to those bees, nor even to Hymenoptera. Ants, yellow jackets, hornets and their kin are eusocial. Termites (Isoptera) have even more complex, eusocial, colonies than do the bees.

Honeybees are insects of the Order Hymenoptera, Superfamily Apoidea, Family **Apidae**, Tribe Apini that contains only one Genus *Apis*. Within the genus are several species discussed in greater detail in Chapter VI – 25. All species are characterized by building comb from the bees' wax (Chapter II – 7) that they secrete from glands on the underside of their abdomens. Other bees also construct comb in their nests, but a

Section I – Bees from the Outside: General External Anatomy, Body Segment by Body Segment
Chapter 1
Introducing Bees: Their Place in the World of Life, General Anatomy, Exoskeleton & Some Basic Concepts

variety of other anatomical and behavioural characteristics distinguishes the honeybees from other bees.

Figure 1-3. The three castes of honeybee: queen, worker, and drone.

At this point, it is worth a short detour into the general concept of **sociality in insects**. Sociality has various grades from aggregation to true sociality. Some insects and various species of bees nest in aggregations, each female having her own nest. Other insects swarm together, as do many flies (Order Diptera) including midges. Monarch butterflies (Order Lepidoptera) spend their winter in aggregations in Mexico and elsewhere. Some insects group band together for feeding (e.g., some sawflies [Order Hymenoptera], caterpillars [Order Lepidoptera]) or for migration (e.g., locusts [Order Orthoptera]). Some insects share accommodations. Tent-caterpillars (Order Lepidoptera) are notorious for that habit. Some adult bees also live together, with several mature, fertile females sharing a nest. It is thought that this latter sort of association is how more advanced forms of sociality arose (Chapter V – 20).

Eusocial (truly social) insects are those that live in colonies and have a rigid social structure with division of labour and different castes, as noted above (e.g., Apidae, hornets and their relatives [Family Vespidae], ants [Family Formicidae] and termites [Order Isoptera]). The social group is often referred to as a **colony**. If the colony persists from one generation to the next (i.e., daughter colonies are formed while the mother colony is still active), the highest level of eusociality has been reached. Honeybees make an excellent example of true eusociality.

The colony lives communally in a **nest**. The nest is the framework built by, and around which, the colony operates to rear its young and to reproduce. The nests of hornets and yellow jackets are made of the paper they make from wood fibre. Termite nests may be earthen mounds or made of carton, or combinations of both, created by the colony. Honeybees make their nest of waxen comb and, depending on the species, this may be within a cavity or out in the open (Chapter V – 19). Stingless bees nest in cavities above or below ground. Their nests are complex structures of separate and distinctive brood-rearing parts and food-storage parts (see Chapter V – 19).

Section I – Bees from the Outside: General External Anatomy, Body Segment by Body Segment
Chapter 1
Introducing Bees: Their Place in the World of Life, General Anatomy, Exoskeleton & Some Basic Concepts

The term **"hive"** refers to the domicile for a colony of honeybees (cavity-nesting species) (Chapters VI – 24 & 25) or bumblebees (Chapters VI – 25 & XI – 47). The term is sometimes extended to aggregated artificial domiciles of some managed solitary bees (e.g., leafcutter bees and orchard bees) (Chapter XI – 47).

Some General Remarks on Terminology

When one is learning a new topic, or refreshing and updating existing knowledge, it is useful to place the information into a broad context, and to understand some of the standard and basic terminology. Bee biology and apiculture has a large and specialized vocabulary. Some comes from biology, some from entomology, and some is special to the study of bees and practice of beekeeping. One may ask, "How much detail needs to be understood to be a competent apiculturalist or beekeeper?" The answer is not easy to give. It depends on the level of expertise the student wants to achieve.

In this book, the amount of detail given is far more than is needed for basic beekeeping. In fact, a student that learns everything in this book would have mastered a great deal of basic biology, basic entomology, basic bee biology, and a great deal about ecology and evolution. Knowing those basics helps one to understand the nature of bees and the reasons for using various management practices in beekeeping. Students using this book should not try to memorize the details presented, but should strive to become familiar enough with the overall content through becoming conversant with the terminology that is clearly important. Students may ask "how does one know what is important and what is not?" Again, the answer is not easy, but is the same. It depends on the level of expertise the student wants to achieve. Nevertheless, some basic rules apply.

1.   Highly specialized terms that are used in only a few places and contexts should be recognized well enough to be looked up as and when needed.
2.   Specialized terms that arise many times comprise the important vocabulary of the topic. The meanings of those terms, and the contexts to which they apply, should become familiar enough to be understood when read or heard, and to be used in writing and conversation.
3.   Concepts presented throughout this book rely on the use of terminology of both types noted above. It is the concepts that students need to understand. Hence, as a student, one must address personal expectations about the depth and breadth of knowledge sought.

Section I – Bees from the Outside: General External Anatomy, Body Segment by Body Segment
Chapter 1
Introducing Bees: Their Place in the World of Life, General Anatomy, Exoskeleton & Some Basic Concepts

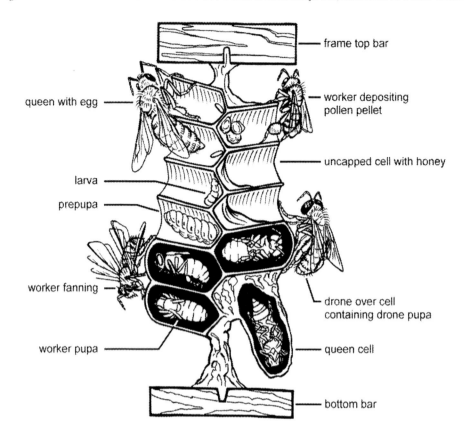

frame top bar

queen with egg

worker depositing pollen pellet

uncapped cell with honey

larva

prepupa

worker fanning

drone over cell containing drone pupa

worker pupa

queen cell

bottom bar

Figure 1-4. Looking at a cut section of comb in the brood chamber of a hive of honey bees. The labels refer to some of the items and subjects that are presented in this book. The "top bar" and "bottom bar" are described in Section VIII. The developmental stages and the activities of the bees depicted in this drawing are described Section IV.

CHAPTER I – 2

THE HEAD:
THE SENSORY & FEEDING CENTRE

What Is the Head?

The heads of almost all advanced animals are centres of both feeding and the senses (Figure 2-1). Supported on the head are the **mouthparts** through which foods pass into the foremost part of the alimentary tract or gut. The sensory organs of the head are the most complex of the entire body and occupy much of the surface. **Optical** (vision and light perception), **chemical** (smell and taste), and **tactile senses** are all represented by organs of the head.   In many animals, including bees, **auditory** sensitivity (hearing) is also located on the head. The proximity of these organs to the brain provides for almost immediate processing of sensory information and reactions on the part of the animal (see Section VII).

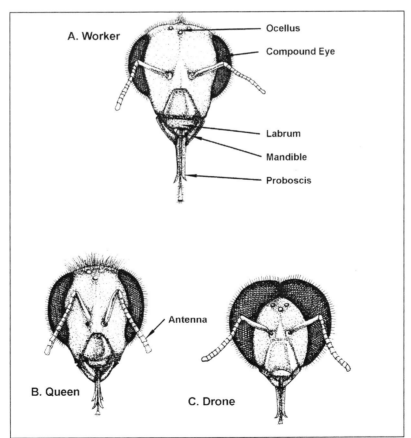

Figure 2-1. The heads of the three castes of honeybees. Workers (**A**) have the longest mouthparts, and the drones (**C**) have the largest compound eyes and longest antennae. The queen's (**B**) compound eyes are slightly smaller, relative to body size, than those of workers.

The most conspicuous sensory feature is the pair of **compound eyes**. These function as photoreceptors and are the main organs for the detection of colour, shape, and movement (Chapter VII – 26). Also present on the top of the head are three smaller light sensors, the **ocelli**. The **antennae** function in both chemical and tactile modes (topochemical) and are involved in auditory perception (hearing). The mouthparts are highly sensitive to chemical stimulation (i.e., taste) as well.

The head also carries other structural and sensory features that are less conspicuous, but are nonetheless important to an understanding of the form and function of insect and honeybee anatomy.

All insects have much the same general anatomy of the head. The mouthparts differ markedly from group to group, but the same parts can be found in a huge diversity of forms. Between different insects there are differences in the antennae, eyes, ocelli, and so on, but the general theme is common to all insects. General entomology books show the diversity of structures. Even within species of honeybees, the main features of the head of honeybees are different in each caste. The **compound eyes** are huge in drones, but smaller in both workers and queens. The **antennae** are longer in drones. Drones have 12 flagellar segments, but workers and queens have 11. The **mouthparts** are largest in workers, much smaller in queens, and even smaller in drones.

The Five Eyes of Two kinds

The **compound eyes** of insects are composed of a few to several thousand facets known as **ommatidia** (ommatidium in the singular). In honeybees there are estimated to be about 4–5,000 ommatidia per compound eye in the worker and 7–8,000 in the drone. The queen's compound eyes are a little smaller than those of the worker, with about 1,000 fewer ommatidia.

Their function is light sensing and gathering. Light enters each facet through the corneal lens into the cone lens to be focused onto the retinula cells. Retinula cells with their visual pigments, as in vertebrates, stimulate and generate nervous impulses to the brain. Each facet may be involved in one or all of four main tasks in sensing optical information (explained in more detail in Chapter VII – 26) .

1) Perceiving the plane of polarized light
2) Recognizing patterns
3) Detecting motion
4) Seeing in colour—Ultraviolet is a primary colour to insects, along with blue and green, but most insects cannot see red (red is black to most insects).

In young bees, there are hairs between the ommatidia. They are shed as bees age. They are not connected to nerves, and their function is unknown.

Between the compound eyes are three **ocelli.** They function as photoreceptors and secondary organs for sensitivity to light. In contrast to the compound eyes, they detect only light intensity; no image is possible. They respond to dim light, which the compound eyes cannot (Chapter VII – 26).

Feeling, Tasting, and Smelling: Multi-tasking by the Feelers

On the front of the head is a pair of **antennae (feelers)**, which are segmented and freely moving (Figure 2-2). Each antenna has three major regions: the **scape (scapus), which** is the first apparent segment that attaches the antenna to the insect head, the **pedicel (pedicellus)**, which is the second apparent segment, and the remaining segments, which compose the **flagellum**.

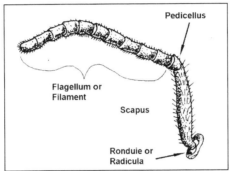

Figure 2-2. Left antenna of a worker honeybee.

The antennae are multi-purpose appendages. They are used as chemoreceptors (Chapter VII – 27), mechanoreceptors (Chapter VII – 29), and as auditory sensors (Chapter VII – 28). By using the chemoreceptors in their antenna, honeybees can sense chemicals in the air (smell) or on a surface (taste); determine the temperature around them; and detect CO_2 concentrations and even humidity in the air (Chapter VII – 27).

Mechanosensory hairs measure the position of the antennae relative to the head and the position of the pedicel relative to the scape (Chapter VII – 29). The degree to which the antennae are deflected by wind may be used to measure the speed of flight. The fine sensory hairs on the tip of the antennae are also used for touch, sensing texture (Chapter VII – 29) and smell.

The antenna is also an auditory sense organ. Inside the antennal pedicel is located the bee's ear, or **Johnston's organ**. It is composed of a ring of mechanosensory cells (scolopidia) that sense movement of the intersegmental membrane between the pedicel and flagellum. The movement of this membrane is induced by sound waves deflecting the flagellum (Chapter VII – 28).

Feeding by Chewing and Sucking: Jaws That Swing Sideways & Proboscis as a Complex 2-way Double Straw

Under the head are the **mouthparts**, indeed a complex array of appendages. They include organs for ingestion of food, chewing, and chemoreception (taste; see Chapter VII – 27).

The mouthparts are composed of **mandibles** (jaws) (Figure 2-3) and a series of segmented parts making up the **proboscis**, a two-way feeding and tasting straw (Figure 2-4). The food canal starts from the tip of the proboscis and leads up into a channel between the bases of the paired maxilla and the labium. The proboscis in honeybees is about 6.5 mm long, but varies according to the race of honeybee (see Chapter VI – 24). It can be exceedingly long (longer than the whole body) in some species of bees (e.g., the euglossines; Chapter VI – 25). When not in use, it is folded and tucked into a groove (**fossa**) under the head.

The mandibles are largest in the queen, but most specialized in the worker (for wax manipulation, feeding on pollen, grasping other bees' mouthparts, etc.). They are least developed in drones.

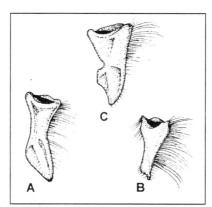

Figure 2-3. The mandibles of a worker honeybee (**A**), drone (**B**), and queen (**C**).

Behind the mandibles is a pair of segmented **maxilla**. Those enfold the segmented **labium** which is made up of two lateral **labial palpi** (palpus in the singular) and a median **glossa** (tongue). The palpi enfold the glossa. The glossa itself is tubular and encloses the salivary canal. Thus, a cross-section of the whole **proboscis** shows an outer sheath of the maxillae (on the front) combined with the labial palpi (behind) comprising the food canal and enclosing the tubular glossa that extends beyond the reach of the maxillae and palpi. Taste (chemoreception) is located mostly on the tip of the glossa (**flabellum**), tips of labial palpi, and on minute maxillary palpi. The hairs on the tip of the proboscis are especially long and, apart from their sensory roles, they are also used in lapping up nectar.

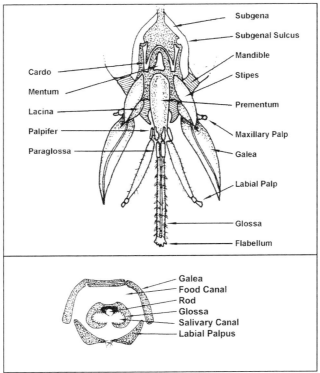

Figure 2-4. *Above:* the proboscis and the sucking apparatus of a worker honeybee: the mandibles are not shown. *Below:* cross-section of proboscis.

Inside the Head: What Goes on in There?

The head contains a large array of internal features despite its small size. Most of these are considered in greater detail in other parts of this book, but it is useful to introduce them here to place them in the perspective of the whole anatomy of the bee. Within the head is **(1)** the brain and associated nerves, **(2)** organs, muscles, and glands associated with feeding and food, **(3)** muscles associated with the mouthparts and antennae, **(4)** air bladders (**tracheal sacs**) that allow oxygen to reach the organs within the head, and **(5)** internal bracing that adds strength to the head capsule.

1. The brain and associated nerves are discussed in greater detail in Chapter III – 9. In brief, **the brain** consists of the supraoesophageal ganglion + circumoesophageal commissures + suboesophageal ganglion. The nerves of the **brain** innervate the compound eyes, the ocelli, the antennae, the mouthparts, and the ventral, double nerve trunks that extend the length of the body.

2. The organs, muscles, and glands associated with feeding and food make up an intricate assembly of interacting components.

The **mouth** is the entry to the **alimentary tract**. It is between and behind the mandibles, and in front of the base of the labium. Food taken through the mouth is

15

ingested by the action of the **cibarial** or **sucking pump.** This organ is a highly muscular part of the alimentary tract that allows sucking and "swallowing" by closing off the mouth and pushing the food into the first part of the oesophagus (Figure 2-5). The **oesophagus** is a pipe that conveys food from the head, through the thorax, and into the anterior part of the abdomen (Chapter I – 3).

The ducts of the hypopharyngeal glands, salivary glands, and mandibular glands all end in the vicinity of the mouth. Their secretions are delivered to the mouthparts by peristaltic pumping muscles, such as are shown in Figure 2-5 for the **salivary syringe.** The ducts of the hypopharyngeal glands open into the mouth. Their secretions are used for feeding larvae and the queen (Chapters II – 7 & IV – 13). The salivary secretions are used in preliminary digestion of food (salivary glands: Chapter II – 7), and the mandibular gland, associated with the mandibles, produces secretions used in communication (Chapter V – 16).

The **hypopharyngeal glands** themselves are located in the head. These glands are discussed in more detail in Chapter II – 7 and vary in their levels of activity according to the age of worker honeybees (Chapter IV – 15).

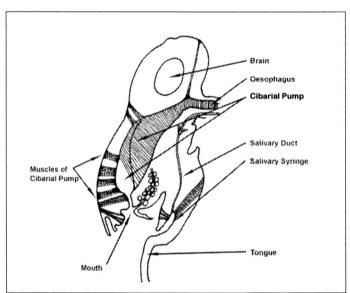

Figure 2-5. Diagram showing some internal features of a honeybee's head in longitudinal section. Note the cibarial pump, salivary syringe, and associated structures.

3. Muscles contract and relax. Only during contraction do they exert force, either by pulling or by squeezing (e.g., peristaltic pumping muscles). The **muscles** that operate the mandibles and the proboscis, and those that allow movement of the scape of the antennae, all operate by pulling and relaxing to varying amounts. To illustrate the principle of pulling and relaxing, the mandibular musculature serves well (Figure 2-6). When the abductor muscle contracts and the adductor muscle relaxes, the jaw swings open. When the contraction and relaxation is reversed, the jaw swings closed to bite.

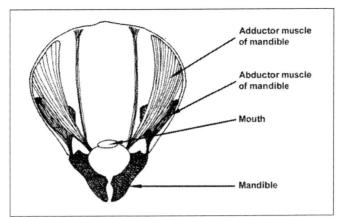

Figure 2-6. Diagram showing some internal features of a honeybee's head in transverse, vertical section (face-on view) with special attention paid to the mandibular muscles and their attachment to the mandibles.

4. The air bladder (tracheal sac) that allows oxygen to reach the organs within the head is thin-walled. Air reaches it from the thorax, and through fine, but reinforced, pipes, atmospheric oxygen reaches the brain, sensory organs, muscles, and glands within the head (Chapter III – 10).

5. Internal bracing. When the muscles of the mouthparts, antennae and sucking pump are working, why doesn't the head collapse? Within the head is an internal bracing of chitin, the **tentorium** (Figure 2-7). The location of the tentorium can be seen by microscopic examination of the front of the head. Pits between the compound eyes and below the antennae lie on a visible sutures. Those pits mark the spots beneath which, inside the head, the tentorium extends to the back of the head capsule, where similar pits can be found by careful inspection. Those minute structures are used in identifying wild bees.

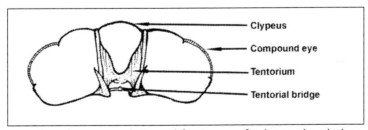

Figure 2-7. Diagram showing some internal features of a honeybee's head in cross-section with special attention paid to the tentorium. The tentorial pits are not labeled.

The Back of the Head

At the back of the head capsule is the **foramen magnum** (a hole) through which passes the **alimentary canal**, **nerve cords**, **trachea** (air tubes), and **dorsal aorta** to and from the **thorax**. A membranous, flexible and elastic chitinous, tubular connection allows for movement of the head on the thorax, and links both together. "Neck" muscles control the movement and position of the head with respect to the thorax. Those

17

muscles, along with the oesophagus (Chapter II – 5), the dorsal aorta (Chapter II – 6), and ventral nerve cords (Chapter III – 9) thread through the elastic and flexible chitinous tube making up the neck. The head also gets some support from a special pointed extension of the front of the first segment of the thorax (**prothorax**). Special long hairs on hair plates that pointed extension brush on the back of the head capsule and so feel the relative positions of the head and thorax. Bees sense the position (**proprioception**) of their heads relative to the rest of their bodies (Chapter VII – 29) by nervous information arriving at the brain *via* the thorax.

CHAPTER I – 3

THE THORAX:
THE LOCOMOTORY CENTRE

What is the thorax? The thorax of an insect is the second major body segment. It is the locomotory centre, bearing three pairs of legs and two pairs of wings, as in most insects. The insect thorax is, itself, made up of three segments, each with a corresponding pair of legs, and the hinder two each with a pair of wings. The thorax and its locomotory appendages is a marvel of natural engineering.

The three thoracic segments are:
I. **Prothorax** with the first pair of legs (prothoracic legs)
II. **Mesothorax** (mid segment and largest) with the second pair of legs (mesothoracic legs) and first pair of wings (mesothoracic wings)
III. **Metathorax** (last segment) with the third pair of legs (metathoracic legs and second pair of wings (metathoracic wings)

The segments with wings can be referred to as a unit called the **pterothorax** (meso- + metathoracic segments). The final part, the **propodeum,** looks as if it should be part of the thorax, but is really part of the abdomen shifted forward to cover the hind part of the thorax proper (Figure 3-1).

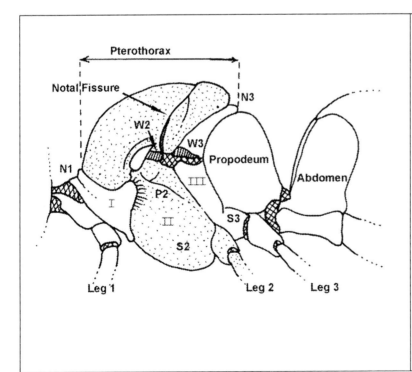

Figure 3-1. Diagram of the organization of thoracic segments of a bee: I = prothorax , II = mesothorax, III = metathorax (II + III = pterothorax). The notum is the back or dorsal plate; three notal segments are pro, meso, and metanotum (as N1, N2 [not labeled, but see Figure 3-5], and N3); the sternum is the ventral plate (S2, S3 are meso and metasternum); the side plate is the pleuron (P2, the mesopleuron, is labeled). The meso- and metathoracic wing bases are shown (W2 and W3; there is no prothoracic wing). The bases of the pro-, meso-, and metathoracic legs are shown.

To understand the marvelous intricacies of how insects walk, use their legs for various tasks, and fly requires a deeper examination of the thorax (Figure 3-1). The thoracic segments are divided into plates of chitin with elastic membranes in between from top to bottom. The **notum** is the back of the bee (its segments are the **pronotum, mesonotum, metanotum**). The **pleuron** (sides) and **sternum** (venter) make up the walls and underneath of the thorax on each side. The **legs** are inserted on the sterna of each thoracic segment. The **wings** are inserted between nota and pleura of meso and metathorax (the middle of the bees' "side"). Remarkable slits, notably the notal fissure and that for insertion of the wings on each side, are crucial parts of the flight mechanism (described below).

The **legs** of insects are more complex than our own with only four segments from thigh to toes. Insect legs are made up of five major segments. In general, the first segment, attached firmly to the sternum of each thoracic segment, is the **coxa** (small and conical). The second segment is the small, cylindrical **trochanter**. The third is the long, more or less cylindrical, **femur**. In bees, specialization in the legs can be noted as they extend away from the axis of the body. The fourth segment is the long **tibia**. The form of the tibia differs depending on the type of bee, the thoracic segment with which the leg is associated, and the bee's sex. The fifth segment of the leg is the highly flexible foot, or **tarsus**, itself segmented and taking on different forms of specialization (Figure 3-2). The last small segment of each leg, sometimes considered to be a special sixth segment, forms a complex "toe".

Within the bee's tarsus there are remarkable specializations. It is divided into 6 subsegments, or **tarsomeres**. The first is the long, cylindrical and most specialized segment, the **basitarsus**. Beyond that are 5 short tarsomeres. On the last segment, known as the **pretarsus**, there are two tarsal claws and an inflatable pad known as an **arolium**.

One can think of the joints as follows: an "inflexible hip" between the thoracic sternum and the coxa, another, but flexible, "hip" between the coxa and trochanter, a "knee" between the trochanter and femur, another "knee" between the femur and tibia, an "ankle" between the tibia and basitarsus, and a series of "knuckles" within the tarsus.

Muscles connect from segment to segment to allow for movement (similar to the arrangement described for the jaws in Chapter 2), and tendons flex the claws in the pretarsus.

The **fore and hind legs** of worker bees are highly specialized. The **foreleg** (prothoracic leg) has an **antenna cleaner** at the "ankle". The antenna cleaner comprises a notch with comb-like array of hairs (setae) in the upper part of the basitarsus and a hinged flange at the end of the tibia. To clean the antenna, the bee raises its forelegs, alternately, over the antenna. When the leg is extended, this "ankle", which works more like an "elbow", opens the notch in the basitarsus, and the antenna flips into it. As the "ankle-elbow" is closed with the flexing of the joint, the antenna is caught in the notch by the special flange of the end of the tibia. The bee then lowers its leg in the flexed position and the antenna is so drawn through the cleaner and the dust, dirt, and pollen are combed off.

The **middle leg** (mesothoracic leg) is not specialized.

The **hind leg** (metathoracic leg) is highly specialized for pollen collecting and packing. The specialization is again at the same "ankle" as is the antenna cleaner on the foreleg.

It is important to understand the structure of the hind legs of worker bees to appreciate how they function in the collection of microscopic pollen grains that are the main source of protein for bees. The tibia of the hind leg is highly modified to form **a pollen basket (corbicula)** on the outer surface (Figure 3-2 (C)). The tibia is broad and flattened and has long, curved hairs along both edges to form a basket-like structure. Near the end of the tibia, in the mid-line of the basket is a single, long hair which helps anchor the pollen load in the basket. On the inside surface of the basitarsus (Figure 3-2 (D)) is the **pollen brush** made up of rows of stiff, closely packed bristles. The bee grooms pollen from its body using all its legs. The pollen passes backwards until it is brushed up with the pollen brush. Pollen gathers in the brush. When there is enough, the bee rubs the brush of each leg against the **pollen rake** of the tibia of the other leg. That action compresses the pollen into a small mass that is caught on the flat area on top of the basitarsus (called the **auricle**) when the leg is extended. As the leg is flexed, the pollen caught in this "ankle-elbow" is pressed out and up, and into the **corbicula**. There, pollen accumulates into a **pollen pellet** because it is sticky or, if dry, may be moistened by the bee with nectar from the flowers it is visiting. Loaded corbiculae look like large, most often yellow, "saddle-bags" on the hind legs of pollen-collecting bees (Figure 3-2 (C)).

In honeybees, the legs of queens and drones do not show these modifications. In other bees, such as bumblebees, the legs show similar specializations in both workers and queens: both have to gather pollen. In other bees, pollen collection is achieved by other specialized structures. For example, leafcutter bees carry pollen on the underside of the abdomen, which has specialized hairs for the purpose.

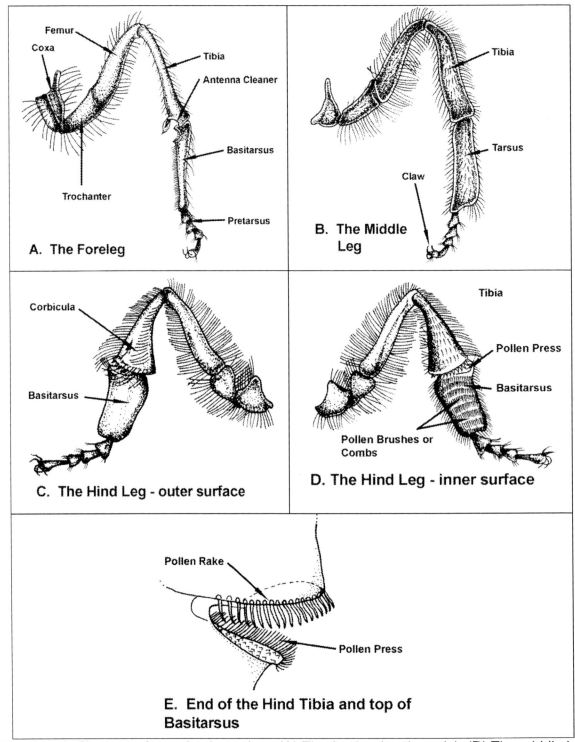

Figure 3-2. The legs of a worker honeybee: (**A**) The foreleg (prothoracic), (**B**) The middle leg (mesothoracic), (**C**) Outer surface of the hind leg (metathoracic), (**D**) Inner surface of the hind leg, (**E**) Enlarged end of the hind tibia with pollen rake.

The wings of insects are made up of two membranes, upper and lower, which are actually outgrowths of the body wall! The **veins** provide mechanical support to the wings. In most insects, the wings are folded back, over the body, when not in use. To fly, the insect extends the wings laterally. The veins at the leading edge of the wings are the most pronounced. In most insects the forewings and hind wings beat together, but are not joined. In bees, though, the pair of wings on each side function as a single wing in flight. They are hooked together by upturned **hamuli** (hooks) on the hind wing (metathoracic) that engage a fold, decurved, on the trailing edge of the forewing (mesothoracic) as the wings are extended for flying.

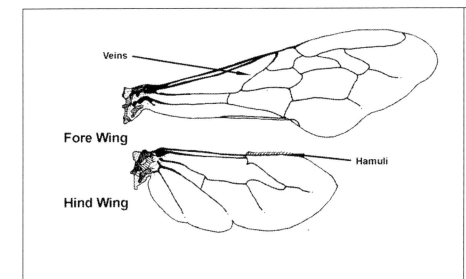

Figure 3-3. Diagram of the wings of a bee. Note the complex of axillary (chitinous) plates at the base of the wing. These provide flexibility and allow for the pitch of the wings to be changed. The hamuli of the hind wing hook onto the fold on the trailing edge of the forewing so that the wings act together as a single airfoil in flight.

Flight in insects is another marvel of natural engineering and a mechanical, aerodynamic bag of tricks. In cross-section, the surface of the wing is not flat, but is convoluted. The shape traps a boundary layer of air that becomes part of the wing to give it an aerodynamic shape. The total effect is to produce a thick leading edge, a convex upper surface, and concave lower surface that tapers to the back. This is similar to the wings of birds and aircraft and works on the same principle. Air flowing past the wing has farther to travel over it than under it. Thus "lift" is generated proportionate to the air speed past the wing and the airplane, bird, or insect can take off.

Although extending wings in strong winds can give lift and flight (as in gliders or sail planes), animal flight usually requires power. Birds and bats flap their wings, human beings use engines. Insects flap their wings, too, but in a most remarkable way.

The **flight mechanism** in insects uses two types of muscles, the **direct** and **indirect**. The direct flight muscles attach to the wing bases, but the indirect flight muscles are so called because they do not attach directly to the wings. Even so, the indirect flight muscles provide the power for flight. How can that be?

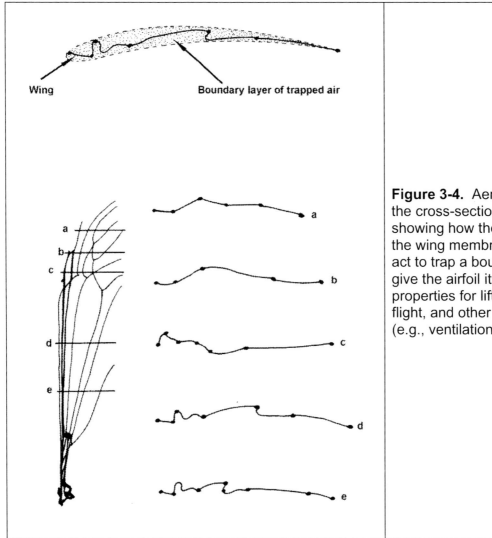

Figure 3-4. Aerodynamic layer and the cross-sections of the wing showing how the convolutions of the wing membrane and wing veins act to trap a boundary layer of air to give the airfoil its aerodynamic properties for lift and thrust during flight, and other uses of the wings (e.g., ventilation and fanning).

The **indirect flight muscles** are of two types; **dorso-ventral** and **longitudinal**. The dorso-ventral muscles connect the inside of the notal plates with the inside of the pleural and sternal plates of the thorax. The longitudinal muscles connect the front and back of the pterothorax by attachments to special parts (apodemes) at either end.

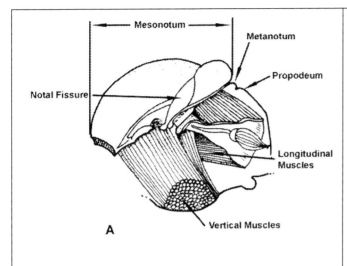

Figure 3-5. Longitudinal view of the inside of the pterothorax of the mesonotum (N2) and metanotum (N3) showing organization of longitudinal and vertical (dorso-ventral) indirect flight muscles.

Both sets of muscles work in concert to provide the power for flight. The fascinating way by which this mechanism works requires the elasticity and flexibility within the exoskeleton of the pterothorax. The **notum** and **pleuron** have an area of elastic cuticle between them. Into that longitudinal slit are inserted the wing bases. The **mesonotum** has a "split" of elastic cuticle, making a dorso-ventral slit called the **notal fissure**. It is the combination of longitudinal and dorso-ventral flexibility upon which the indirect flight muscles work to power the wings. The mechanism is fascinating.

The indirect flight muscles pull on the thorax. Longitudinal muscles pull front to back. Dorso-ventral muscles pull top to bottom (notum to sternum). When the longitudinal muscles contract, the dorso-ventral slit closes, but the longitudinal slit opens. The notum and sternum move apart. When the dorso-ventral muscles contract, the longitudinal slit closes and the dorso-ventral slit opens. Finally, the notum and sternum come together.

Remember that the wings are inserted in the longitudinal slit. The lid (notum) and bottom (**pleuro-sternum**) function like a box, with the notum slipping into the pleuro-sternum by a small amount. Thus, as the notum is pulled down and slightly into the pleuro-sternum by the contraction of the dorso-ventral muscles, the wing bases are pushed down. But they also hinge on the top of the bottom part of the longitudinal slit (i.e., on the top of the pleuron). Thus, as the wing-bases are pushed downward over the inside of the pleuron, the wings are flipped upward outside the body of the bee. That provides the upstroke of the wing. The dorso-ventral muscles are the wing elevators.

Next, the **longitudinal muscles** contract as the dorso-ventral muscles relax, and the ends of the box are pulled together, closing the notal slit (**notal fissure**), opening the longitudinal slit, causing the notum and pleuro-sternum to move apart, drawing the "lid" out of the "box-bottom", and pulling up on the wing bases proximally. But distally, the wing bases are attached to the top of the pleuron, so the wings beyond the body of the thorax are flipped down. Thus, the downstroke of the wing is created. The longitudinal muscles are the wing depressors.

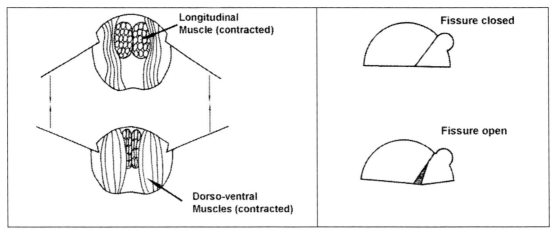

Figure 3-6. The indirect wing muscles and flight mechanism. **Left:** Cross-section of thorax showing action of the longitudinal and dorso-ventral muscles. **Right:** Lateral view of corresponding states of the thoracic tergites (dorsal plates).

The thoracic mechanism has two points of stability: one when the wings are fully up, and the other when they are fully down. Little muscular force is needed to flip the wing system from one stable position to the other. That's because the wing bases roll on the top edge of the pleuro-sternum. The mechanism is not unlike one of those "party favours" (Figure 3-7) that click as you squeeze the "jaws" together and click again as they spring apart when pressure is released. These clickers consist of a rigid, curved top to which a piece of spring metal is attached at one end. The piece of spring metal has two stable positions, one when held up toward the top by pressure and the other, a down position, when no pressure is being applied. It requires large amounts of energy to keep the piece of spring metal in an intermediate position and is very difficult to control.

Then the cycle starts again, with the result that the wings flap up and down. However, this does not explain the fast (200 cycles/second) wing beating.

Rapid wing beating comes about through a complex interaction of the physics of the thoracic cuticle and anatomy, and the physiology of muscular contraction and relaxation. The muscular contractions and relaxations initiate a rhythm that is amplified by the physics of the thoracic structure as it starts to vibrate harmonically with the frequency of the muscle activity rhythm.

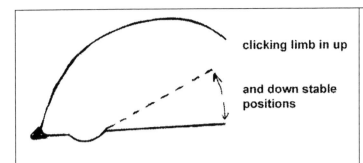

clicking limb in up

and down stable
positions

Figure 3-7. Party clicker, in longitudinal section, showing arched, rigid top, and spring metal clicking limb in up and down stable positions. The click is made as the spring metal changes position with a snap.

Merely flapping the wings up and down does not produce forward motion. Forward motion is needed for the aerodynamic shape of the wings to provide lift, and for the bee to go anywhere. During the up and down motion of the wings, more complex forward and backward strokes are executed. The tip of the wings describes a figure "8". On the downstroke, the wings also move forward and twist downward at the front. On the upstroke, the wings move back and reverse their twist. This combination makes the wings work to lift the bee and act somewhat like an oar and "row" it forward through the air.

The **direct flight muscles** are small and attached to the complex of fine **sclerites** (Figure 3-8; also described as axillary plates, see Figure 3-3) that flex between each other in a way similar to the bones of our wrist and hands) at the base of the wing. These muscles control the fine-tuning of the wing's pitch or attitude, allowing the leading edge of the wings to be tilted upward or downward on either or both sides of the bee. Such precision movements allow the bee to climb, descend, and steer as it flies in the same way as the flaps on the wing of an airplane are used. When both wings are tipped down by contraction of indirect flight muscles attached to the forward wing bases, then the bee loses altitude. When the bases at the rear of the wings are pulled on, the front of the wings are elevated and the bee gains altitude. If the front of one wing pair is elevated, and the front of the other depressed, the bee turns.

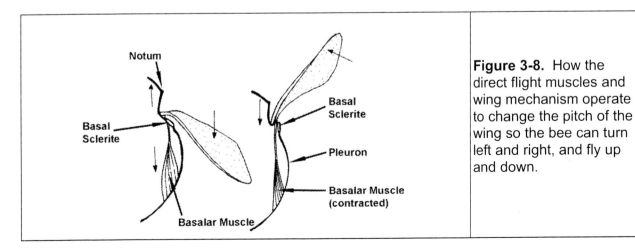

Figure 3-8. How the direct flight muscles and wing mechanism operate to change the pitch of the wing so the bee can turn left and right, and fly up and down.

27

Other Features of the Thorax

The **spiracles** are the breathing apparatus of the bee. There is a large one on each side between the prothorax and mesothorax covered by a flap of cuticle of the prothorax. The spiracles are discussed in more detail as part of the Respiratory System (Chapter III – 10), and the large thoracic spiracles become important in the biology of the honeybee's parasitic **tracheal mites** (Chapter IX – 34). There is also a small spiracle in the deep membranous fold between the upper ends of the meso- and metathorax. Within the thorax is a complex of air sacs that are part of the respiratory system (Chapter III – 10).

The **oesophagus**, part of the alimentary tract (Chapter II – 5), runs straight through the thorax between the indirect flight muscles and into the abdomen. The **dorsal aorta,** a major component of the circulatory system (Chapter II – 6) runs from the abdomen, arching through the thorax to the head. In the thorax, the dorsal aorta has some special features that are probably involved in heat exchange and perhaps in cooling highly energy-consumptive flight muscles. The mechanisms also involve the haemolymph (or blood) returning from the head to the abdomen (but see Chapter II – 6 on the circulatory system and Chapter V – 19 on thermoregulation for details).

Also running through the thorax are the paired **ventral nerve cords.** They have large and complex ganglia in the thorax. They are involved in the regulation and coordination (from one side of the body to the other and between segments of the thorax) of locomotion by walking and flight. Their structure is described in more detail in Chapter III – 9.

The **thoracic glands** are located in the ventral part of the front of the thorax. They are important in various aspects of the life of bees, and insects in general as described in Chapter II – 7 on the glandular system and Chapter IV – 14 on metamorphosis.

The **propodeum** (see Figure 3-5) is really the first segment of the abdomen. It contains compressible air sacs, aiding in the process of pumping air during respiration (Chapter III – 10) and provides for space for the movement of true **thoracic apodemes** or **phragma** (the special places on the inside of the exoskeleton onto which muscles attach).

CHAPTER I – 4

THE ABDOMEN:
THE FACTORY WALLS

The abdomen of a worker is made up of 10 segments, the first being the **propodeum** (it appears to be part of the thorax (Chapter I – 3). The next is the **petiole**, or "wasp-waist". The nine remaining segments appear as six. Thus, abdominal segments II, III, IV, V, VI, and VII are visible. Segments VIII, IX, and X are concealed within segment VI. They are highly modified, reduced, with parts missing so as to be barely recognizable as segments (Figure 4-1).

Each complete segment has a **tergum** and a sternum. The tergum consists of the large plates making up the back of the bee's abdomen while the sternum is made up of the smaller plates of the underbelly of the bee's abdomen. Between the terga and the sterna are membranes of elastic cuticle that allow for dorso-ventral expansion and contraction. Between the segments are **intersegmental membranes**, also of elastic cuticle. These allow for longitudinal elongation and contraction. The dorso-ventral and longitudinal muscles that connect between tergum and the sternum, and between the segments, power the movements of the abdomen. Such movement is important in many activities. During breathing, the abdomen pumps air in and out of the spiracles and into the respiratory system (Chapter III – 10). While stinging, the female bee curves its abdomen downward to deploy the sting. The drone similarly curves its abdomen downward during mating. When dancing to communicate (Chapter V – 17), the abdomen may be waggled from side to side, or up and down.

Segments II to VIII have a **spiracle** (entry port for breathing) on each side in the lower part of the tergum. The spiracles have tiny muscles that allow them to be opened and closed, and are equipped within with fine hairs that filter the air. Surrounding the spiracles on the outside are larger, plumose body hairs that probably also have some filtering and protective roles.

Segments IV to VII have paired **wax glands** on each sternal plate (Figure 4-2 (B)). The glandular tissue of columnar cells is actually inside the cuticle, but the cells secrete their wax to the outside. Thus, the cuticular surface of the wax gland area appears shiny and is called the **wax mirror**. The wax (Chapter II – 7), as it is secreted, forms small, thin flakes, or scales, which are clearly visible to the naked eye. Segments III to VI have rearward extensions that overlap the wax glands. The **wax scales** form here and are groomed off by the bee for building comb.

Another prominent gland that secretes to the outside of the abdomen is the **Nasonov scent gland** (Figure 4-2 (A)). The glandular area is actually the anterior part of the tergum of segment AS VII. About 500 individual secretory cells have individual microscopic pores that conduct their secretions to the outside. Normally, the Nasonov gland is not visible, but when bees fan (Chapters V – 16 & 19) at the entrance of the hive, they may tip the point of the abdomen downward to expose the gland. The gland liberates odour signals as part of the repertoire of chemical communication among honeybees (Chapter V – 16).

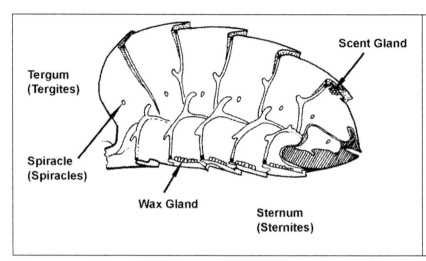

Figure 4-1. The internal view of the abdominal wall of a worker honeybee illustrating organization of the abdominal segments and other features. The scent gland is the Nasonov gland.

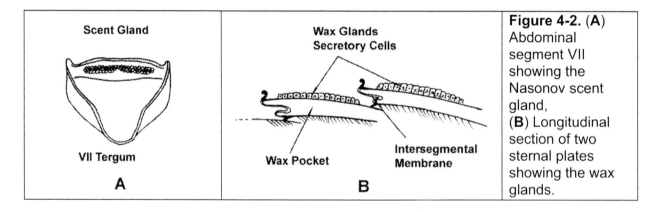

Figure 4-2. (A) Abdominal segment VII showing the Nasonov scent gland, **(B)** Longitudinal section of two sternal plates showing the wax glands.

At the end of the abdomen is the highly complex **ovipositor/sting** apparatus in the queens and workers, or the male genitalia of the drone (Chapter III – 11). The sting apparatus consists of a lever system made up of a complex of hinged chitinous plates, with attached muscles, a sheath that covers the actual stinger, and the sting itself. Normally, the sting is contained in the sting chamber. The point of the sting may be slightly exposed or not. When the bee starts to sting, the complex of plates rotates, thrusting the sting itself out of the chamber and into the victim. The entire coordination of stinging is through neuromuscular reflexes involving the last abdominal ganglion (Chapter III – 9).

The **sting** (Figure 4-3) of queen and worker honeybees is a modified ovipositor (egg-laying organ) that has been remodeled solely for injection of poison. The egg-laying function has been lost and the sting is held out of the way of the egg as it is laid by passing to the outside via the **vagina** (Chapter III – 11) to the bottom of a comb cell. The bee's sting is a multi-purpose organ, depending on the caste. **Workers** use their stings in defence of the colony. **Queens** use their stings in offence when attacking their sister queens in the colony during the process of queen replacement following swarming. This is during the **battle royal** (Chapter V – 20).

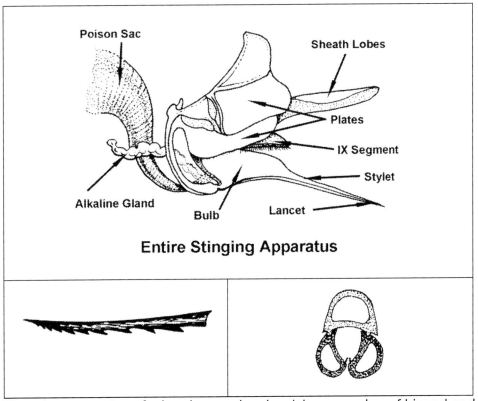

Figure 4-3. The sting apparatus of a bee is complex, involving a number of hinged and rotating plates that push the sting, with its stylet and lancets, out and into the victim. **Below left:** stinging lancet with barbs. **Below right:** cross-section of sting showing stylet above and two lancets: the venom channel is formed by all three parts.

The **shaft** of the sting is a complex organ comprising a dorsal **stylet** and paired ventral **lancets** (Figure 4-3). The lancets slide back and forth against each other, supported by the stylet. The three parts enclose the **poison** or **venom canal** (Figure 4-3). The lancets in the worker are straight and barbed, like a harpoon head, at the end; the lancets of the queen have barbs that are much less developed than those on the worker's sting. Thus, when the worker's sting penetrates skin, it cannot be withdrawn and the worker is doomed. She leaves the sting and its associated internal organs stuck in the victim when she flies off. The queen's sting is curved and relatively smooth, allowing the queen to sting, withdraw, and sting again, as she must to survive the battle royal. The stings of other bees and wasps are feebly barbed or smooth (e.g., bumblebees, carpenter bees, hornets, and yellow jackets). In some parasitic wasps, like ichneumons, the egg passes down the long, slender ovipositor (made up of the same parts as in bees), which is also the sting that injects poison to immobilize the prey as the egg is laid within the latter's body.

The **venom** (its composition is discussed in Chapter X – 43) is produced by the poison gland and is stored in the **venom (poison) sac** (Figure 4-3). An **alkaline gland** (Figure 4-3; also Chapters II – 7 & V – 16), also associated with the sting apparatus, produces a secretion which may lubricate parts of the sting apparatus.

Also at the end of the abdomen is the **anus,** a small cone of chitin, on a highly modified segment X.

Within the abdomen are major internal organs. Those are discussed in the following chapters:
- The whole of the Reproductive System (Chapter III – 11)
- The whole of the Excretory System (which is associated with the Digestive System) (Chapter II – 5)
- Most of the Alimentary Tract and Digestive System (Chapter II – 5)
- Much of the Respiratory System (Chapter III – 10)
- Much of the Circulatory System (Chapter II – 6)
- Some of the Nervous System (Chapter III – 9)
- Some of the Endocrine and Exocrine Systems (Chapter II – 7).

SECTION II

Bees' Innards: Internal Organs for a Healthy Bee
Processing Food & Waste, Circulation, Glands Galore, & Fat

Now that the structure of an insect's body is somewhat familiar from the outside, and aspects of the workings of the exoskeleton and musculature are understood, the "guts" beckon. Again, the internal organ systems of bees are similar to those of most insects. And they carry out the same bodily functions as do organ systems in people.

The **alimentary** or **digestive tract** extends from the mouthparts on the head to the **anus** on the tip of the abdomen. The anterior parts are specialized for ingestion of liquids and solids. One part is the **honey stomach**, a special extensible bladder for harvesting nectar. It is in the anterior part of the abdomen. The main digestive part of the gut, called the **ventriculus**, occupies much of the interior of the abdomen. Linked to it, as it joins the hinder parts of the intestine, are the equivalent of kidneys. Lots of **Malpighian tubules** float through the blood, better-called **haemolymph**, cleaning up the chemical waste products of metabolism. Those tubules pass their waste to the anterior intestine where it joins the indigestible remains of food to be passed back to the **rectum** and out through the **anus**. The honeybee's rectum is remarkably capacious, being able to store a winter's worth of waste at once.

Bees do store nutrients within their bodies. **Fat** is important in wax secretion, in making eggs, and as stored energy.

The **circulatory system** has no real veins and arteries. It consists of a single pulsing tube, the **heart**, in the top of the abdomen. The blood enters that tube through special valves, and is pumped forward through the thorax and into the head in a **dorsal aorta**. The **haemolymph** gushes out to bathe the brain and other organs in the head. Then it flows back in a carefully directed fashion to the abdomen, bathing the internal organs as it does so. In the abdomen, the haemolymph bathes the ventriculus, picks up nutrients, and is cleansed by the **Malpighian tubules**. The head gets the richest, cleanest blood. The haemolymph is not involved in oxygen transport.

The **glandular system** of insects, like our own, is dispersed throughout the body. Some glands secrete to the outside (exocrine glands), and others to the inside (endocrine glands). The **exocrine glands** have several functions, including production of bees' wax, and of signal chemicals recognized by other bees. The **endocrine secretions** regulate growth, development, metabolism, water balance, and activity. Some of the endocrine secretions originate in the nervous tissue of the brain and nervous system.

CHAPTER II - 5

PROCESSING FOOD & WASTE:
THE ALIMENTARY TRACT & EXCRETORY SYSTEM

The **alimentary tract** and **digestive system**, or **gut**, extends from the **mouth** to the **anus** as a tube running the complete length of the bee (Figure 5-1). Associated with it is the **excretory system** that cleanses the blood (better termed haemolymph; Chapter II – 6)) and conducts the accumulated chemical wastes to the gut for elimination from the body.

As part of the head, the **mouthparts** (Chapter I – 2) make a complex and sealable tube for sucking up liquids, and also comprise the chewing mandibles that are involved in processing solid food. In bees it is difficult to describe the actual **mouth** as a single organ of ingestion. In the area where the mouthparts articulate with the head capsule is a complex **buccal** or **preoral cavity** that receives food from the **food canal** within the **proboscis**, or from the **mandibles**. The entry to the **alimentary tract** is located in the buccal cavity and can be closed by strong muscular contraction of the **cibarial Pump** (**cibarium**). The mouth is closed when the **sucking pump** activates, thus food is pumped into the **oesophagus**, a tube extending through the head and **thorax**. There is no clear demarcation between the cibarium and the oesophagus. Also associated with the buccal cavity are the ducts from various glands, including the **hypopharyngeal glands** (Chapter II – 7). There are also two sets of **salivary glands**, one in the head and the other in the thorax. Both share a common duct that conveys **saliva** into the buccal cavity where it mixes with the food to dilute, soften, or lubricate it as it is ingested. The saliva can also be pumped out through the proboscis to dissolve food such as crystalline or highly sticky syrups (nectar) (Chapter II – 7). The hypopharyngeal glands are also thought to produce enzymes involved in food processing in the buccal cavity and honey stomach (see below).

The first part of specialization in the **gut** is the **crop** or **honey stomach** in the anterior part of the **abdomen proper**. When empty, the crop wall is highly folded. It is highly distensible, allowing the crop to function as a storage bladder for nectar and honey. Its maximum capacity is about 70–80 mg of nectar, or about 7 µL (microlitres), amounting to about 85% of the weight of an individual worker bee (Chapter V – 18)!

Extending into honey stomach from behind is a short, muscular organ, the **proventriculus**. The proventriculus acts as a valve, regulating the flow of material from the oesophagus (including the honey stomach) into the **ventriculus**, the main digestive organ (see below), and the reverse (i.e., vomiting). The proventriculus moves about in the honey stomach, somewhat like a stubby tentacle with an X-shaped opening between its thick, bristly and muscular lips. The special hairs on the ends of each of the four lips gather pollen and other particulates from the stored nectar and allow that to be passed back through this "mouth" and into the ventriculus.

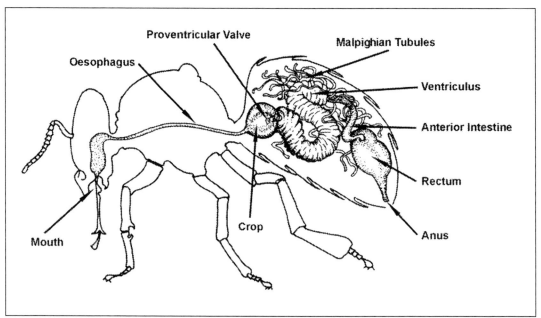

Figure 5-1. The digestive and excretory systems of a worker honeybee.

As food enters the ventriculus, it becomes surrounded by a very thin, fine and permeable membrane of chitin, the **peritrophic membrane.** It is thought that this membrane may offer some protection for the ventricular wall against abrasion, especially by pollen as it passes down the gut. The ventriculus itself is a highly folded and distensible, muscular tube and is the part of the gut in which digestion and absorption of food take place. The inner surface of the ventriculus is lined by a thick layer of columnar **epithelial cells** that secrete enzymes for digestion and absorb digested material in the form of organic molecules. The digestive process in honeybees is not well understood. Apart from enzymes secreted by the ventricular epithelium, there seem to be specialized microorganisms resident in the ventriculus. Those microorganisms may play an important role in digestion (see Chapter IV – 13). Nevertheless, the digestive enzymes in the ventriculus, regardless of source, are involved in the breakdown of essential nutrients, proteins, lipids, and carbohydrates, which are taken up with other nutrients in the ventriculus by the epithelial cells. It is the epithelial cells that transport the dissolved nutrients across the wall of the ventriculus and into the haemolymph (Chapter II – 6). Outside the epithelial layer are **circular muscles** overlain by **longitudinal muscles**. Those muscles allow for contraction and expansion of the ventriculus, and the pumping (peristaltic) actions that propel the digesting food and associated waste posteriorly.

The **anterior intestine** is merely a tube at the end of the ventriculus. Food waste is pumped into it. The intestine expands into the **rectum**. There, water from the accumulated waste may be absorbed by the **rectal glands**. The rectum also serves as a distensible storage bladder for solid (semi-solid) waste, such as pollen exines (the shells of pollen grains). It can come to be highly distended and to occupy much of the abdominal cavity. Bees normally do not defecate in the hive. Overwintering bees store large amounts of feces in the rectum, which can come to occupy a large amount of the abdominal cavity. On fine days in early spring in temperate parts of the world, bees leave the hive in droves on **cleansing flights**. **Mass defecation** is also part of thermoregulation in the giant honeybee (*Apis dorsata*) of tropical Asia (Chapter V – 19). The posterior opening of the tract is the muscular **anus**.

Excretion in insects is mostly carried out by the **Malpighian tubules**. These tubules have a function similar to that of kidneys. There is no separate urinary tract and opening. Where the ventriculus and anterior intestine join, the Malpighian tubules discharge their wastes. There may be 100 or more of these single-cell thick, fine, blind-ended tubules. They extend throughout the abdominal cavity, twisting and winding amongst the other internal organs, and continually being bathed by the blood (haemolymph). These remarkable tubules function, as do kidneys, to filter and cleanse the haemolymph. The one to few cells surrounding the lumen (cavity) of the tubule extract nitrogenous wastes derived from protein metabolism from the haemolymph. Uric acid and urea are the main waste products passed along and into the intestine for elimination. The Malpighian tubules are also involved in maintaining proper balances of water and salts in the insect's body.

CHAPTER II – 6

CIRCULATING NUTRIENTS:
THE CIRCULATORY SYSTEM

The circulatory system of insects is different from our own, but shares similarities. In insects and in humans, a fluid is pumped and flows through the body to convey nutrients to the tissues, and waste products from them. That fluid contains cells that contribute to the protection from disease. However, apart from those broad similarities between vertebrates and insects, there are many differences.

Insect "blood" is mostly clear, yellowish liquid. It is not involved in transporting oxygen around the body, as is the blood of vertebrates. It is more correctly called **haemolymph**. It is composed of liquid plasma in which are suspended cells called **haemocytes**. The number of haemocytes in insect blood may vary widely, depending on conditions and life stages. Several kinds can be recognized by microscopy. Their most important roles are in phagocytosis, the ingestion of foreign particles. Thus, they act like human white blood cells, ingesting invading bacteria, other microorganisms, and tissue debris. They can also have an important role in surrounding, or encapsulating, other larger pathogens or internal parasites. These cells are also active in the clotting of the haemolymph following a wound and, then, in wound healing. Thus they prevent penetration of microorganisms into the body cavity. All in all, they function in the same ways as do similar cells in our own blood: defence against infections and diseases.

The insect circulatory system is **"open"**. There are no veins, arteries, or capillaries. Haemolymph bathes all the organs and muscles of the insect body. Nevertheless, it must flow throughout the body in some sort of orderly manner if it is to function in the distribution of nutrients and the removal of wastes (Figure 6-1).

There is a single, dorsal blood vessel in insects. The posterior part of that vessel, the tubular **heart**, is the main pump that creates the flow of haemolymph. It is closed at the posterior end, and runs the length of the top of the abdominal cavity (segments III to VII), along the mid-line. This long, muscular, tube with 5 pairs of slits (**ostia**) pulses by muscular contractions and relaxations. When the heart increases in diameter (relaxed muscles in diastole), the ostia open and the haemolymph enters. When the heart contracts in systole, the ostial valves close and haemolymph is pumped forward. In the petiole between the abdomen and thorax, there is a curious wiggly section of this single blood vessel. That section may be important in heat exchange for thermoregulation (Chapter V – 19). The continuation of the vessel in the petiole and thorax is the **dorsal aorta**. Haemolymph is pumped through the arching aorta between the indirect flight muscles in the thorax and into the head. The flexing of the indirect flight muscles may aid in the pumping of the haemolymph forward during flight (Chapter I –3) or for thermoregulation (Chapter V –19).

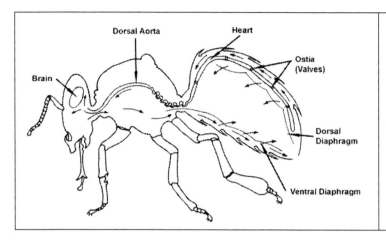

Figure 6-1. The circulatory system of a worker honeybee.

The vessel passes through the neck and into the head, where the haemolymph gushes out, bathing the brain and all other structures on and in the head (antennae, eyes, mouthparts, muscles, glands, nerves, cibarium, and oesophagus). It then washes backward, through the foramen magnum, into the neck and thorax, where it bathes the thoracic muscles and nervous system, and flows down the legs into the wing veins (at least during eclosion from pupa to adult (Chapter IV – 14)).

As the haemolymph passes through the petiole, it tends to become confined to the ventral part of the anterior abdomen by the thin, yet muscular, **ventral diaphragm**. The ventral diaphragm extends from the mesothorax to abdominal segment VII. It lies above the **ventral nerve cords**. This diaphragm is quite muscular and its beating causes haemolymph to flow backward. The haemolymph, having bathed the nerve cords, sloshes out into the abdominal cavity, bathing all the organs and absorbing nutrients passing from the digestive system by diffusion (Chapter II – 5). At this time, the haemolymph is in intimate association with the many Malpighian tubules (Chapter II – 5) and can be cleansed of the metabolites gathered from the passage through the head and thorax.

The nutrient-enriched and cleansed haemolymph flows upward and over the internal organs of the abdomen, passing upward and over the edges of the **dorsal diaphragm**. That diaphragm is a thin, weakly muscular organ attached segmentally to the top of the insides of the terga from abdominal segments III to VII that forms a **pericardial cavity** through which the haemolymph re-enters the heart.

Thus, the richest, cleanest haemolymph is pumped straight to the head, losing its nutrients to the tissue of the brain, head, thorax, and ventral nerve cords as it courses back into the abdomen while gathering metabolic wastes. The haemolymph *does not* carry oxygen as our blood does. However, the haemolymph is important in the transport of water soluble carbon dioxide as the waste product of respiration (Chapter III – 10).

CHAPTER II – 7

GLANDS GALORE:
THE EXOCRINE & ENDOCRINE SYSTEMS

Glands are secretory organs and they can be placed into two broad categories. **Exocrine glands** secrete to the outside of an organism whereas **endocrine gland** secretions stay within the organism. Glands may secrete nutrients, enzymes, hormones (regulators of metabolism and growth (Chapter IV – 14)) or semiochemicals (signal chemicals) (Chapter V – 16), especially pheromones (chemicals that influence other members of the same species).

The main **exocrine glands** of bees have been mentioned in previous chapters as being associated with particular body segments and parts (Figure 7-1). There are also exocrine glands distributed generally over the body surface. Little is known about the full role that the secretions play in the life of bees, but some of the secretions are important in assuring the integrity of the outside of the cuticle, providing waterproofing oils and waxes to reduce water loss through evaporation, and adding to the complex blend of pheromonal chemicals secreted by more conspicuous organs. Glandular activity over the entire body is important during moulting (Chapter IV – 14).

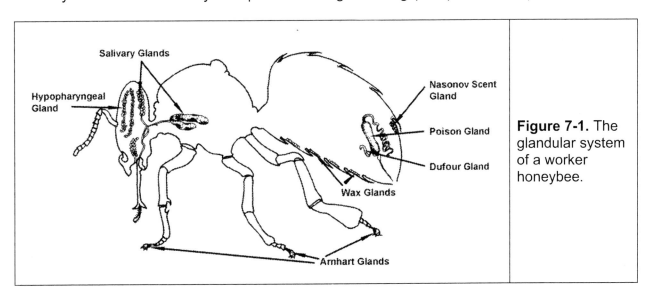

Figure 7-1. The glandular system of a worker honeybee.

In the head are a number of glands that serve various functions in the life of a honeybee. The **hypopharyngeal glands** are important for production of larval food (Chapters IV – 13 & 15) and of enzymes involved in digestion of food (Chapter II – 5). They are long and comprise many secretory bulbs that discharge into a central duct. The whole gland loops around on each side of the inside of the head. They are highly developed in the worker, and change size according to age (in younger bees they are larger) and nutritive status (feeding on pollen, which is nutritionally rich, causes these glands to expand). The secreted food for brood, called **royal jelly** or **worker jelly**, is similarly nutritionally rich in proteins, lipids, and carbohydrates, but otherwise is chemically highly complex (Chapter X – 41). These glands also secrete **invertase, glucose oxidase**, **diastase** and other enzymes involved in digestion of nectar as it passes into the honey stomach.

The **postcerebral glands** are also paired and in the head. They consist of clusters of secretory bulbs that discharge to a central duct. They secrete a fatty, oily substance in workers and queens. These secretions likely have salivary and lubricative functions. Secretions from these glands can intermingle with those from the thoracic glands, which also secrete saliva (see below).

The **mandibular glands** are paired glands associated with the mandibles. In the worker bee, they secrete a variety of compounds including fatty substances associated with brood food and pheromones associated with alarm signals (Chapter V – 16). In the queen, they secrete an array of pheromonally important compounds (**queen mandibular pheromone** or **QMP**); as many as 24 have been identified (Chapter V – 16). In the drone, they secrete a drone-to-drone attracting pheromone that may be used by honeybees and other bees to mark congregation areas (Chapters V – 16 & VI – 21).

There are also a few other small glands in the head, but their functions are unknown. The **thorax** and the **feet** house yet more exocrine glands.

The **thoracic salivary glands** (Figure 7-2) are paired and lie in the ventral part of the thorax. They are derived from the silk glands that are present in the larvae (see below, and Chapter IV – 12). Their clusters of finger-like secretory bulbs discharge into ducts that conduct the saliva to a reservoir at the anterior of the gland. The secreted, watery saliva is used for diluting and dissolving sugar. The secretions can intermingle with those from the postcerebral glands.

The **Arnhart glands** or **tarsal glands** are associated with the last tarsal segment of the leg (Chapter I – 3) of all three castes, but are best understood in workers. These glands, a single layer of cells lining a cuticular sac that occupies much of the interior of the last tarsal segment, secrete a variety of chemicals making a pheromone blend. It is not known how the chemical secretions cross the wall of cuticular chitin to accumulate in the sac. They are attractive and possibly, in bumblebees, also repellent. The blend is sometimes called the "footprint pheromone" (Chapter V – 16) as it is assumed that the chemicals are deposited as the bee walks.

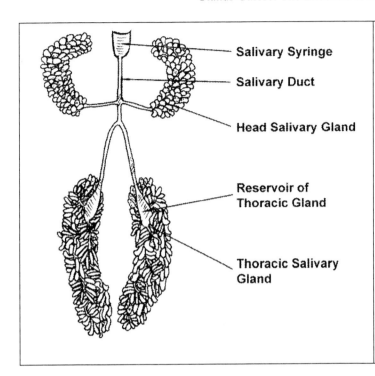

Figure 7-2. Salivary glands of the head and thorax of a worker bee.
(Redrawn from Snodgrass, 1992.)

The exocrine glands of the abdomen also deliver pheromonal and non-pheromonal chemicals. For some, it is difficult to separate the two functions. However, the **Nasonov gland** secretes a blend of at least 7 chemicals as signals to other honeybees (Chapter V – 16). This gland is a single area on the top of the abdomen on the anterior part of the tergum of abdominal segment VII. The glandular cells number about 500 and have individual microscopic pores that conduct their secretions to the outside.   Worker bees expose this well-developed gland by stretching the terga apart as they tip the point of the abdomen downward. Then, by fanning their wings, they disperse the pheromone into the air as part of the repertoire of chemical communication among bees. It is attractive to other worker bees, and seems recognizable to them as different from hive to hive (Chapter V – 16).   There are other, smaller glandular areas associated with abdominal tergites III, IV, and V that seem to secrete pheromones involved in worker recognition of the queen and perhaps in attracting drones to queens.

The **wax glands**, of which there are eight all together (two per segment) secrete **bees' wax**, the material from which comb is constructed. Segments IV to VII have paired wax glands on each sternal plate. The glandular tissue of columnar cells is actually inside the cuticle, but the cells secrete their wax to the outside, into the **wax pocket**, created by extensions of the sternites of segments III to VI which overlap the wax glands. Thus, the cuticular surface of the wax gland area appears shiny and is called the **wax mirror**. The wax (Chapter X – 39), as it is secreted, forms small, thin flakes, or scales, which are clearly visible to the naked eye. It is from the hind edges of the wax pockets that the **wax scales** protrude and are groomed off by the bee for building comb. It has been suggested that bees' wax also has a pheromonal role (Chapter V – 16).

An array of glands is associated with the ovipositor and stinging apparatus. The **poison gland** is the largest gland associated with the sting in workers and queens. It comprises two, long secretory tubules that unite to form a single duct that drains into the much larger poison sac. The nature of the venom is described in Chapter X – 43. The **alkaline** or **Dufour's gland** also directs its secretions to the sting chamber. Its secretions may be as any or all of the following: as a lubricant during stinging or egg laying, or both, as a waxy covering for eggs, perhaps attaching eggs to the floor of the cells, and possibly as a queen-produced egg-marking pheromone. **Koschevnikov's gland** is associated with the complex of plates of the sting apparatus. The secretion from the large glandular cells accumulates among special hairs that become exposed when the sting is deployed. Among worker bees, the secretion is part of the blend of alarm pheromones, but the secretion of this gland from a newly mated queen is highly attractive to workers (Chapter V – 16). Other parts of the sting apparatus have been reported to have similar secretory areas and to be possibly part of the production of "alarm pheromone" that elicits defensive behaviour (Chapter V – 16).

In the queen there is a **spermathecal gland**. It is small, consisting of two tubes that open into the spermathecal duct (Chapter III – 11). It produces polysaccharides and proteins, presumably as nutrients for sperm stored in the spermatheca (Chapter III – 11).

There are glands associated with the drone's **reproductive system** (Chapter III – 11). The paired, and relatively huge, **mucous glands** secrete mucus, which makes up part of the semen. The mucus' function is presumably nutritive for the sperm and lubricative during mating (Chapters III – 11, VI - 21). Small, **accessory glands** are also associated with the male reproductive system, but their function is not known.

As mentioned above, in the larvae are paired **silk glands**. These long, convoluted tubes secrete silk as the final larval stages go to pupation. The silk is secreted through paired openings on the larval mouthparts and is used to make an almost imperceptible cocoon within the cell (Chapter IV – 12). These glands become the thoracic salivary glands in adult bees.

Of course, there are secretory cells within the wall of the ventriculus. A special ring of cells just after the proventriculus secretes the peritrophic membrane (Chapter II – 5), and secretory cells produce the various enzymes involved in digestion (Chapter II – 5).

The **endocrine** glands are generally inconspicuous, and most endocrine secretions are produced by patches of cells that are not aggregated into recognizable organs. Within the brain and other parts of the nervous system (especially in the ganglia) (Chapter III – 9) are patches of secretory cells whose products eventually diffuse into the haemolymph to be distributed throughout the insect's body (Figure 7-3). Thus, nervous, neurosecretory, endocrine, and circulatory functions are all enmeshed in a complex, neurohaemal web that regulates growth, maturation, differentiation, moulting, and behaviour (Chapters IV – 14 & 15).

In general, there are two patches of **neurosecretory cells** in each side of the brain. Their secretions pass down the axons of the nerve cells to two small organs, the **corpora cardiaca** (corpus cardiacum in the singular), on the dorsal aorta. Those are connected to another pair of small organs, the **corpora allata** (corpus allatum in the singular), that lie on each side of the oesophagus. These in turn are connected to the ventral part of the brain, the suboesophageal ganglion. Thus, the neurosecretions from the brain can be conducted throughout the nervous system. Chemical and physiological interrelations take place in this system, in which the brain's neurosecretory products stimulate the corpora cardiaca to produce a hormone that is carried in the haemolymph to the **prothoracic glands**. These glands are a paired, diffuse patch of cells that, when stimulated, produce moulting hormone (ecdysone) that diffuses into the haemolymph and becomes dispersed throughout the insect's body, thereby coordinating moulting and growth. The secretions of the corpora cardiaca also stimulate the corpora allata to produce juvenile hormone. That complex of interactions is described in more detail in Chapter IV – 14.

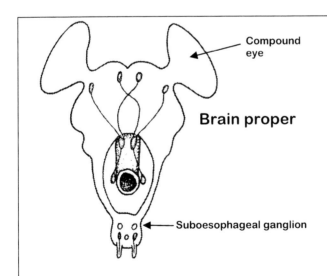

Figure 7-3. Schematic view of an insect's brain showing the positions of the Compund eye, neurosecretory cells, the corpora cardiaca (dorsal aorta not shown), the corpora allata, neurosecretory cells in the suboesophageal ganglion. See also Chapter IV – 14; Figure 14-2.

CHAPTER II – 8

FAT BODY & NUTRIENT STORAGE

Many animals accumulate fat as a means of storing energy. For some, it allows for hibernation, in others it permits long-distance migration, and for all, it can be used when food is scarce.

The **fat body** of bees is the storage organ for energy reserves in the form of fat (**lipids**) and carbohydrates (**glycogen**). The fat body is made up of a layer of cream-coloured, ovoid cells on the roof and floor of the abdominal cavity. The cells also contain protein reserves (such as albumin). As larvae feed and grow, they accumulate fat, which they can carry through to adulthood. Correspondingly, as adult bees age, the size of the fat body declines. Old workers mostly show little sign of having fat body.

Fat body cells can be found in close association with actively secreting cells, such as the wax glands and Nasonov gland (Chapter II – 7).

Glycogen, as a short-term storage carbohydrate, is also found in the muscles and intestinal tissues. Some energy is stored in the haemolymph as a special insect sugar, **trehalose**. This sugar is synthesized in the fat body and released into the haemolymph where it is readily hydrolyzed to produce glucose, the high-energy sugar for fueling the muscles in flight, thermoregulation, and for general bodily activity. Glucose and other sugars are often present in the haemolymph as a result of digestion and diffusion from the gut.

Fat body has also been implicated in the magnetic sense of bees (Chapter VII – 30).

SECTION III

Bees' Innards: Internal Organs for an Active Bee
The Nervous System, Respiration, & Reproduction

Insects, especially bees, have a remarkably astute appreciation of their surroundings. The external sensory array (Chapter I – 1) has an amazing capacity to capture information, but more is to come on that (SECTION VII). However, for that information to be useful to a bee, or any animal for that matter, the sensory array must be coupled to a central processing system. That is the **central nervous system** (**CNS**). Although there are some general similarities between the CNS of human beings and insects, there are profound differences. The insect **brain** is shaped like a malformed doughnut: the gut runs through the hole. There is a major lobe, sometimes called the brain proper, over the alimentary tract. Nervous tracts, called commissures, that extend from above and around the gut on each side, join a large ganglion under the gut. Major nerves extend to the elements of the sensor array of the head from the upper and lower parts of the brain. The paired nerve trunks to the rest of the body run parallel and posteriorly, but with large, nerve-linking ganglia associated more or less with each segment of the thorax and abdomen. Those **segmental ganglia** coordinate reflexes involved in flight and fight.

The respiratory system is unique to insects. Segmentally arranged ports, or **spiracles**, allow air to be pumped into and out of the body. In bees, the air enters large, balloon-like sacs before it is conducted throughout the body in a complex of smaller and smaller ramifying tubes, the **tracheae**, right to the organs needing oxygen. The whole body, and especially the abdomen, is involved in pumping the air. At the end of the finest tracheal tubes, the oxygen is finally transferred into the awaiting tissue. Respiratory wastes, such as carbon dioxide and water, become incorporated into the haemolymph.

The **reproductive system** of insects shows some bizarre structures. Insect mating behaviour can also be very strange. The arrangement of the reproductive organs of female bees shows similarities with most animals, in the presence of **ovaries, oviducts, vagina**, and **associated glands**. In honeybees, though, the queen has a **spermatheca** in which she keeps a supply of sperm good for several years of laying eggs. When a mature, fertile queen is laying at her peak, she produces over 1,000 eggs a day. Her ovaries are so large that she is unable to fly (Chapters V – 20 & VI – 21). The drone's reproductive apparatus qualifies it for being among the most bizarre in the insect world. The presence of **testes, vas**, and **associated glands** is standard, although the structures are somewhat unusual in form. But the complex of structures of the **intromittent organ, endophallus**, or **penis** is reminiscent of the "Alien Empire". It literally, and audibly if one is close enough, explodes when a drone mates with a queen. When he is to become a "Pop", he pops.

Section III – Bees' Innards: Internal Organs for an Active Bee
Chapter 9
The Nervous System from Stem to Stern: Form & Function

CHAPTER III – 9

THE NERVOUS SYSTEM FROM STEM TO STERN: FORM & FUNCTION

All animals must be able to sense what is around them, and react in appropriate ways to ensure survival. The sensory information (Section VII) and the ensuing reactions are coordinated by the nervous system.

This chapter starts with a description of the bee's entire nervous system, from head to tail. The total nervous system comprises the brain and associated structures in the head, and then a twinned set of ventral nerve trunks with more or less segmentally arranged "mini-brains" or ganglia, and the nerves that run to the muscles and from the sense organs. This chapter discusses briefly the roles of those parts of the nervous system in conveying sensory information from outside the nervous system, and actionary (motor) impulses to the muscles. The means by which nervous impulses travel from sensory origins and for muscular control is complex, but a brief description of the electrochemical processes along the nerve cells as they "fire" and transmit impulses is given. Nervous impulses must jump from nerve cell to nerve cell, via synapses, during their journeys so the biochemical processes are briefly described. It is those synaptic processes that become compromised during neural poisoning, the main mode of action of insecticides.

Section VII on sensory physiology and Chapter V – 18 on foraging expand on aspects of this whole chapter.

What Comprises the Central Nervous System?

The **central nervous system** (**CNS**) in insects comprises the brain and the main nerve trunks that extend the length of the body (Figure 9-1). Thus, it is similar to our own CNS with our brain and spinal column. However, it is different in various important ways. Apart from differences in the way the brain is structured, the main nerve trunks are ventral (not dorsal as in our own bodies), paired (not single as in ourselves), and interconnected, more or less segmentally, with nerve masses called ganglia (ganglion is the singular).

Section III – Bees' Innards: Internal Organs for an Active Bee
Chapter 9
The Nervous System from Stem to Stern: Form & Function

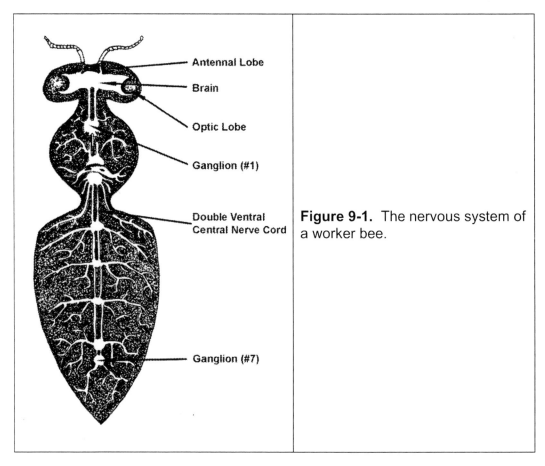

Antennal Lobe

Brain

Optic Lobe

Ganglion (#1)

Double Ventral
Central Nerve Cord

Ganglion (#7)

Figure 9-1. The nervous system of a worker bee.

What Is the Brain of an Insect Like?

The brain of insects (Figure 9-2) comprises the **cerebrum**, or brain proper, and the **suboesophageal ganglion**. The former resides in the top of the head as the main mass of nervous tissue of the CNS. It is estimated that the brain of the honeybee contains only about 850,000 cells. The cerebral mass is thought to comprise three areas, the **protocerebrum, deutocerebrum**, and **tritocerebrum** inherited through evolution from worm-like insect ancestors. These parts are rather indistinct, especially in advanced insects such as bees.

The whole brain is complex and bilaterally symmetrical (Figures 9-1 & 9-2). As studies continue, more and more is learned about the structure and function of the various recognizable areas of the brain. Particularly well studied are the so-called **mushroom bodies** or **corpora pedunculata** (corpus pedunculatum in the singular). As the name suggests, they resemble mushrooms or, perhaps more so, double-headed chanterelles. They are known to be involved in integrating sensory information as it pours in from the general central nervous system, the compound eyes, ocelli, antennae, mouthparts, limbs, and the body surface (Figure 9-3). Moreover, these mushroom

Section III – Bees' Innards: Internal Organs for an Active Bee
Chapter 9
The Nervous System from Stem to Stern: Form & Function

bodies are thought to be centres of memory and learning and change in subtle ways as adult bees age and change their behaviours accordingly.

Thus, the insect brain has the equivalent of left and right hemispheres, as does our own. Just as our brain has central connecting structures (the corpus callosum and associated structure) the insect brain has the **central body**. It seems to be involved in the association of information between the two sides of the brain. This structure appeared to have evolved for the first time in Arthropoda (see Chapter I – 1). It is noteworthy, and indicative of the requirements of complex learning and social behaviour, that the relative sizes of the corpora pedunculata and central body (CP:CB ratio) is the highest in social insects, especially bees, and less in all non-social insects.

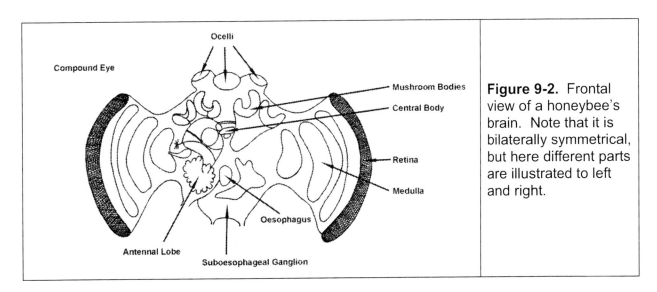

Figure 9-2. Frontal view of a honeybee's brain. Note that it is bilaterally symmetrical, but here different parts are illustrated to left and right.

The peripheral parts of the brain are also bilaterally arranged. The largest parts are the **optic lobes** that include the compound eyes. The **compound eyes** can be considered as part of the brain. The outermost reaches of the brain are the retinula cells (Chapter VI – 26) that receive light, segregate signals of different wavelengths (colours), and allow for visual acuity involved in the perception of shape and size, colour vision, perception of contrast, and stereoscopic appreciation.

Beneath the retina are bands of nervous tissue that are presumably involved in integrating the information gathered by the retinula cells (see Chapter VII – 26 on colour opponency coding and colour vision) and feeding that information into the central reaches of the brain. The three **ocelli** can also be thought of as three dorsal lobes of the brain that end in photoreceptive cells (Chapter VII – 26).

Section III – Bees' Innards: Internal Organs for an Active Bee
Chapter 9
The Nervous System from Stem to Stern: Form & Function

On each side of the frontal part of the brain are the **antennal lobes** into which sensory nervous impulses feed from the antennae and from which nerves extend to allow antennal movements (antennation) (Chapters VII – 27 & 29).

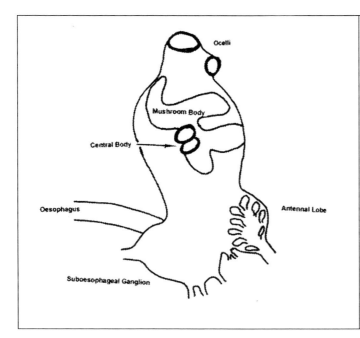

Figure 9-3. Diagrammatic lateral view of the brain of an insect. Note the location of the alimentary tract (oesophagus) and suboesophageal ganglion with one of the two ventral nerve trunks extending posteriorly and the three main nerves to mandible, maxilla, and labium (from front to back) extending down. The large antennal lobe and antennal nerve extend forward. On top of the brain is one of the lateral ocelli, and below that the median ocellus. The complex mushroom body is shown, with the bilobed central body associated with it in the middle of the brain. The compound eye and optic nerve are not represented.

Also, within the cerebral mass are some special neurosecretory cells. These are cells that produce biochemicals, called hormones, that are involved in the regulation growth, development, moulting, behaviour, and so on, in ways similar to the secretions from pineal and pituitary glands in human beings. The location and functions of these cells and the hormones they secrete are presented in Chapters II – 7 and IV – 14.

From the underside of the dorsal part of the brain arise two massive nerve trunks, the **circumoesophageal commissures** or **connectives** pass on either side of the oesophagus. They connect to the **suboesophageal ganglion** (actually the fusion of three large ganglia associated with the segments of the head as represented by the mouthparts). The nerves to and from the mouthparts enter and leave the brain via this ganglion.

The dorsal mass of the cerebrum (including the optic and antennal lobes) plus the circumoesophageal connectives and suboesophageal ganglion are sometimes considered to comprise the whole brain. From the suboesophageal ganglion, **two ventral nerve cords** extend posteriorly through the foramen magnum (Chapter I – 2) and neck, and into the thorax.

In the thorax are more or less segmentally arranged **ganglia**. The first is associated with the prothorax (**prothoracic ganglion**). The next is a large, fused ganglion associated with the meso- and metathorax and propodeum. This ganglion is associated especially with the integration of muscular contractions in flight (see Chapter

Section III – Bees' Innards: Internal Organs for an Active Bee
Chapter 9
The Nervous System from Stem to Stern: Form & Function

I – 3), and coordination of the legs in walking, grooming, and collecting pollen (see Chapter I – 3).

The **double ventral nerve cord** extends beneath the **ventral diaphragm**, through the petiole and into the abdomen proper. There, the ganglia are, again, more or less segmentally arranged. There are five abdominal ganglia in all. The last one is associated with stinging and the muscular coordinatory reflexes involved (see Chapter I – 4). In queen bees it is assumed that that the ganglion is involved in coordination of egg laying (oviposition) (Chapters I – 4 & IV – 12).

What Are Ganglia?

In general, ganglia are masses of nervous tissue, with nerves making interconnections with sensory and motor nerves by way of **associative neurones** (also called neurons). In insects, the ganglia interconnect between the two nerve cords and are site of origin of many nerves. They can be considered similar to telephone switchboards, but much more complex. Messages from and to the brain pass through the ganglia en route to and from all parts of the body. Also within ganglia are nerve shortcuts that allow rapid coordination from side to side of the body, and along its length. The wings must beat in unison and the legs walk and work in harmony (Chapter I – 3).

What Is a Nerve?

A nerve is a bundle of nervous tissue, mostly **axons** (Figure 9-4), making a long or short string. A nerve may extend within the central nervous system (brain and double ventral nerve cord with ganglia), or it may branch away.

Within the nerve are **motor neurones** or **fibres**, which carry information from the central nervous system to the muscles and other ganglia (all the way out to the basitarsus and sting apparatus). These nerves cause the muscles to contract and the limbs and body parts to move. Also, within a nerve are **sensory neurones** or **fibres**. These carry information from the periphery (e.g., the tips and segments of the antennae, the mouthparts and flabellum, the legs and tarsal segments, body wall, etc.) of the body to the central nervous system ganglia and, ultimately, to the brain.

Within the central nervous system and between and within ganglia are **associative neurones** or **fibres** that allow for incoming sensory information (**stimuli**) from the sensory nerves to be connected with, and cause reaction in, the motor nerves, which in turn activate muscles and movement (**reaction**). The neuromuscular circuit is completed.

In ganglia, some circuits are completed as **reflex** connections. Thus, sensory neurones pass information to associative neurones, which in turn pass information to

Section III – Bees' Innards: Internal Organs for an Active Bee
Chapter 9
The Nervous System from Stem to Stern: Form & Function

motor neurones, to complete the sensory neuromuscular circuit all within the ganglion itself and without involving the brain.

Some other circuits must be completed through the brain so the effect of one stimulus can be weighed against the effects of other stimuli, and with the learned experiences of the bee, before a reaction takes place. If learning isn't involved, but the brain is, the reaction can be considered as an **instinct**.

Common reflexes in bees are associated with flying (ganglion 2 in the thorax) and stinging (last ganglion in the abdomen). Bees seem to have instinctive attraction to flowers. Also, bees can learn to associate various stimuli with food (sugar or water; Chapters V – 18 & VII – 27).

Nerve cells or **neurones**, in general, have a cell body from which extend **dendrites** going away from the central nervous system, and an **axon** extending toward the central nervous system. The functions of the nervous system depend on a fundamental property of neurones called **excitability**. Upon being stimulated, a neurone is considered to have "fired". This property involves a change in membrane permeability in response to stimulation.

Neurones, in general, function by electrochemistry, involving the changing of electrical potential across the cell membrane by waves of ionic (**sodium** and **potassium**) changes traveling along a nerve cell's **dendrites** and axon (Figure 9-4).

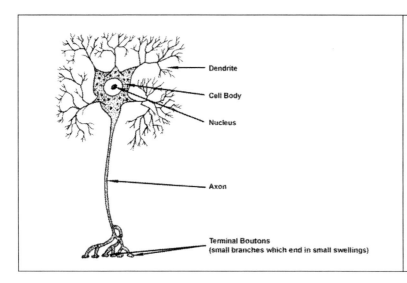

Dendrite

Cell Body

Nucleus

Axon

Terminal Boutons
(small branches which end in small swellings)

Figure 9-4. A neurone (nerve cell). Nerve impulses travel along the axon: from the sense organs towards the cell body in sensory neurones or away from the cell body to the muscles in motor neurones.

How Do Nerve Impulses Travel?

The process of neuronal transmission, or the conducting of nerve impulses, starts when a stimulus activates the neuron. The effect of the stimulus is to cause

Section III – Bees' Innards: Internal Organs for an Active Bee
Chapter 9
The Nervous System from Stem to Stern: Form & Function

sodium ions (Na+) to rush across the nerve cell's membrane into the cell at the point of stimulation. That is called **depolarization**. That initiates a chain reaction whereby Na+ is allowed to rush across the cell's membrane as the impulse travels down the neuron. As the Na+ enters the cell, potassium ions (K+) are repelled from the cell, so that the net chemical electrical charge across the nerve cell's membrane is restored (**repolarization**) in only a millisecond. The cell then re-establishes its **resting potential** (at about –75 millivolts versus the action potential during "firing" of +65 millivolts) by actively pumping the excess Na+ out and allowing the K+ back in. The process of ion exchange along the neuron takes place through special molecular (protein) pores.

What Is a Synapse?

The region of functional contact between two neurones is known as a **synapse**. This represents the point of transfer of nerve impulses. A **synaptic cleft** of about 15 nm occurs between the plasma membranes of the axon of one neurone and a dendrite of the next (Figure 9-5). The direction of impulses defines **presynaptic** and **postsynaptic membrane**. Most synapses are chemical (humoral), but others are electrical.

The chemical synapses involve release of **neurotransmitter** from **synaptic vesicles**. The neurotransmitter diffuses across the **synaptic cleft** to induce excitation in the next neurone of that synapse.

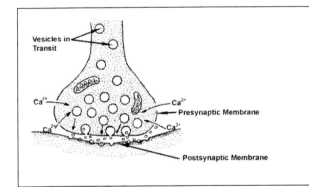

Figure 9-5. Diagrammatic representation of the synaptic cleft. Calcium (Ca2+) also is involved in the depolarization and repolarization processes here.

What Is a Neurotransmitter?

One of the best known neurotransmitters is the chemical **acetylcholine**. Dopamine, octopamine, and hydroxytryptamine, are examples of others that characterize specific kinds of synapses. These powerful chemicals are released across the synaptic cleft and stimulate the next neurone to "fire". If the neurotransmitter chemical remained on the postsynaptic membrane, it would continue to stimulate that nerve. Thus, the synaptic system also produces enzymes to de-activate the neurotransmitter.

Section III – Bees' Innards: Internal Organs for an Active Bee
Chapter 9
The Nervous System from Stem to Stern: Form & Function

The enzymes, such as acetylcholine esterase, quickly destroy the neurotransmitter chemical. Thus, transmission across the synapse ceases and the postsynaptic membrane is not chemically stimulated once the impulse from the presynaptic neurone stops. Nerve poisons, including many insecticides, inhibit the enzymatic activity. Thus, the postsynaptic membrane is overly stimulated, causing nervous exhaustion and death (Chapter IX – 37).

The same set of processes is involved in the transmission of nervous impulses to muscles. The muscles then respond by contracting.

The chemical nature of neurotransmitters and their de-activating enzymes, and the form of synapses, varies in different parts of the nervous system; however, the general principles are the same.

Neuronal transmission is complicated further by neuromodulator chemicals. These are peptides and amines that intensify or reduce normal responses of postsynaptic neurons to neurotransmitters. Serotonin, also known to be involved in human behaviour, may function as a neuromodulator in neuronal transmission in antennae and modulate responses to pheromones (see Chapter V – 16).

CHAPTER III – 10

RESPIRATION: HOW BEES DRAW BREATH

The respiratory system of insects is quite unlike that of human beings. Nevertheless, they breathe by pumping air into and out of their bodies. There are no lungs, and there is no blood to carry oxygen and carbon dioxide. In this chapter, the anatomy of the respiratory apparatus is described, and the method by which oxygen finally makes its way into metabolically active tissues, especially muscles, is introduced. The rates of respiration in honeybees vary widely, depending on their activity, and this aspect of honeybee activity is discussed in Chapter V – 19 on thermoregulation.

What Is Respiration?

During **respiration**, organisms take in oxygen, assimilate it, and use it to oxidize organic compounds in the body. This **metabolism** releases energy used in general activity and generates carbon dioxide. Other products of metabolism, metabolites, are used in the body for activity, growth, storage, repair, maintenance, and reproduction. Still others are wastes that are eliminated via the excretory system (Chapter II – 5).

What Is Insect Blood Like?

Insect "blood", or **haemolymph,** is watery and contains no red blood cells to carry **respiratory gases** (oxygen) in solution, as in human beings. Insect haemolymph is not involved in oxygen transport around the body (see Chapter II – 6), but it transports carbon dioxide in solution throughout the body to be liberated through the tracheal system.

What Is the Route by Which Oxygen Gets to Insects' Body Tissues?

Many small invertebrates respire by direct diffusion of gases through their general skin coverings. Most insects are too large to do that. Also, their integument or cuticle (Chapter I – 1) is too hard, impervious, and waterproof.

For oxygen to enter the insect body, air enters through the **spiracles**. These are exterior openings of the **trachea** situated segmentally along the sides of the body. Thus, they are small breathing pores. Most insects have 10 spiracles, as do bees. Adult bees have three thoracic and seven abdominal spiracles. The first spiracle (associated with the prothorax) is the site of entry of **tracheal mites**, an important pathogen of honeybees (Chapter IX – 34). The third thoracic spiracle is associated with the propodeum, really part of the abdomen (Chapters I – 3 & 4). Abdominal segments II to VIII have a spiracle on each side in the lower part of the tergum. The spiracles have tiny muscles that allow them to be opened and closed, and within are equipped with fine hairs that filter the air. Surrounding the spiracles on the outside are larger, plumose body hairs that probably also have some filtering and protective roles.

The trachea are of cuticle that is continuous with the exterior of the insect exoskeleton, and thus the trachea can be thought of as invaginations of the exoskeleton. They have multiple branches and act as air conduits to the interior. The first part of the tracheal system comprises large diameter **tracheal trunks** of various lengths, depending on the part of the body to which they are directed. In highly active and energetic insects, such as bees, some trachea, especially of the abdomen, are much expanded and thin-walled (Figure 10-1). Here, the tracheal trunks are almost absent, replaced by **tracheal air sacs**. They function as air-storage bags, working in much the same way as the expanded respiratory system of birds, with air storage in their hollow lightweight limb bones. There are two very large tracheal air sacs associated with the abdominal spiracles, one on each side of the abdomen. There are smaller tracheal air sacs in the thorax, head, and even legs.

The thoracic tracheal trunks from the first mesothoracic spiracle is large and long, extending into the head before expanding into an air sac (Figure 10-1).

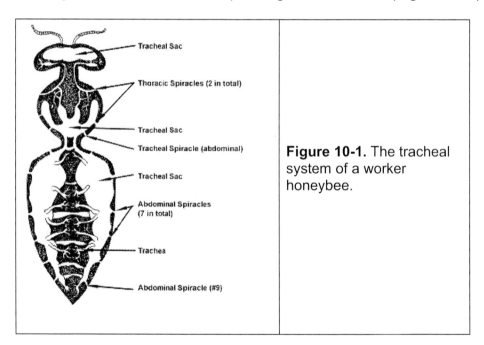

Figure 10-1. The tracheal system of a worker honeybee.

What Do Trachea Look Like and Why Don't They Collapse?

The trachea ramify, dividing into finer and finer tracheoles as they reach into the organs of the body. They are all reinforced with rings of chitin that resemble the reinforcing rings in vacuum cleaner hoses and the like (Figure 10-2). These chitinous rings, called **taenidia,** are lacking from tracheal air sacs; that allows the sacs to collapse partially and expand, pumping air in and out as takes place in bellows or in lungs during inhalation and exhalation. Insects, including bees, engage in active breathing or **ventilation**, which can be seen easily as the active pumping movements of the abdomen when bees are at rest.

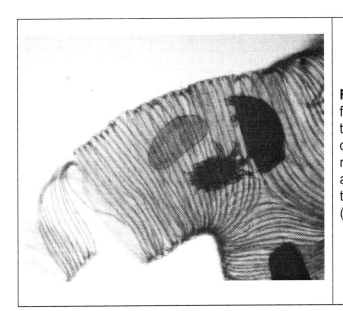

Figure 10-2. The first thoracic tracheal trunk dividing. Note the reinforcing taenidia, and the infecting tracheal mites (Chapter IX – 34).

How Does Oxygen Reach the Cells?

From the air sacs, the trachea ramify throughout the body. They end as fine **tracheoles** and their open, liquid-filled ends abut the walls of the cells, especially muscle, of the body tissue. The liquid absorbs the oxygen from the air within the tracheoles. The liquid is drawn from the end of the tracheole into the underlying cells of the body tissue, muscles, nerves, etc. Thus, the tissue gets oxygen for respiration. After the underlying cell has reduced the oxygen to a minimal level, liquid is returned to the tip of the tracheole to absorb more oxygen.

Although the tracheoles are reinforced against collapse, they do become compressed and deformed as a result of insects' muscular activities. The rapid compression, followed by release and expansion, increases and decreases the pressure differential between the respiring tissue (e.g., muscle) and the gas and liquid in the tracheole. Thus, the gas exchange is not entirely passive, but includes some physiological activity and force. In beetles, frequency of compression and release is about half a second.

How Does Carbon Dioxide Leave the Cells?

Carbon dioxide is 30-fold more easily soluble in water and diffuses from the cells of the body tissue directly into the haemolymph. It exits from the insect's body by diffusion into the trachea and so to the outside during exhalation. Some also diffuses to the outside through softer, more permeable parts of the exoskeleton.

CHAPTER III – 11

REPRODUCTIVE SYSTEMS:
THE ORGANS & HOW THEY WORK

The reproductive systems of honeybees are among the most bizarre in the insect world, involving suicidal and explosive copulation by the males (drones), multiple matings and copious egg production by the females (queens), and optional sterility in workers, which are also female. The anatomy and physiology of reproductive systems of queens, workers, and drones are associated with different functions of these castes in the colony. This chapter starts with the reproductive system of queens, with regard to virginity, mating, and maturation. The reproductive system of workers is similar to that of the queens, but because the workers are usually sterile, there are important differences. The sting apparatus (Chapter I – 4) and its associated glands, including the poison gland (Chapter II – 7), are all part of the reproductive system even though their direct role in reproduction has become lost. The reproductive system of drones is similar to that in most insects, but has some special features that assure transfer of semen to the virgin, or "semi-virgin", queen during copulation.

Figure 11-1. An egg-laying queen, marked with a white dot on her thorax for ease of discovery by beekeepers during hive inspection and monitoring, surrounded by her retinue of workers (Chapter V – 16). Note her long and wide abdomen, packed with ovaries. Her wings are shorter than normal because they have been clipped to prevent her from swarming (Chapter V – 20). That is a common beekeeping practice (Section VIII).

The Queen's Reproductive System

In the mature and reproductively active queen, the paired ovaries occupy much of the interior of the abdomen. They are so large (Figure 11-1) that a mature, egg-laying queen cannot fly, but when the queen first emerges as an adult from her queen cell (Chapter IV – 12), her ovaries are not mature (Figure 11-2 (A)). It takes several days for the features of the mature ovary to become apparent, and only after mating do they become completely mature and functional. The mature **ovaries** (Figure 11-2 (B)) are two large, pear-shaped organs, containing hundreds of closely packed tubular **ovarioles**. Within the ovarioles is a beaded string-like series of developing eggs, **ova** (ovum is the singular), and associated **nutritive cells**. The nutritive cells concentrate nutrients derived from the queen's haemolymph (see also Chapter II – 6) and pass them into the ova where they become **yolk**.

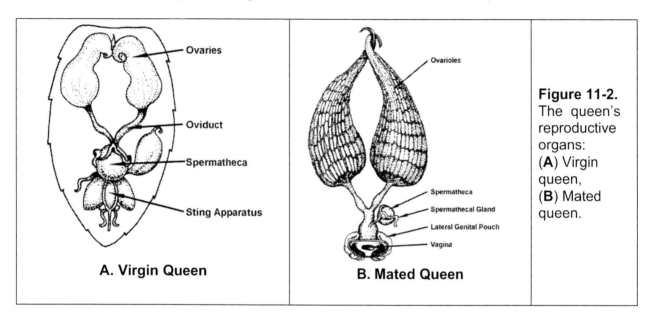

Figure labels: Ovaries, Oviduct, Spermatheca, Sting Apparatus, **A. Virgin Queen**, Ovarioles, Spermatheca, Spermathecal Gland, Lateral Genital Pouch, Vagina, **B. Mated Queen**

Figure 11-2. The queen's reproductive organs: (**A**) Virgin queen, (**B**) Mated queen.

The most mature ova are at the distal end of the ovarioles. As an egg matures, it leaves an ovariole in one or other ovary and passes into the **lateral oviduct** from the ovary. At maturity each egg is slightly curved, 1.6 mm long, 0.4 mm in diameter, and weighs about 0.08 mg (Chapter IV –12).

The lateral oviducts from the left and right join in the short **common oviduct**. The common oviduct is continuous with the larger cavity of the **vagina**. Entering the vagina from the dorsal side is a **spermathecal duct**, leading from the spherical **spermatheca**, in which **sperm** are stored. **Spermathecal glands** (Figure 11-2 (B); Chapter II – 7) are associated with the **spermathecal ducts**. The glands probably secrete nutritive materials to nurture the stored sperm over the years of their potency (Chapter II – 7).

During mating (Chapters V – 20 & VI – 21), which occurs during a short period of the young queen's life and not again, the queen is inseminated with the **semen** of several drones. The semen is first held in the lateral oviducts, but once mating is over, the semen is forced into the **spermatheca**. The semen is stored with its living cargo for the duration of the queen's active reproductive life of several (2–4) years.

As an egg is laid, it passes down the oviduct, through the vagina, maturing finally as it descends. There, spermatozoa are discharged frugally (a few at a time). A single sperm may enter the egg through the **micropyle**, and fertilization can take place (Chapter VI – 22). If the egg passes out of the vagina and is laid without being fertilized, it develops to become a drone. If it is fertilized, it develops into a worker or another queen (Chapter VI – 22).

The Worker's Reproductive System

The ovaries of a worker bee (Figure 11-3 (A)) are reduced to two strands of tissue, vestigial ovarioles, but without ova. The lateral oviducts and common oviduct are present. The spermatheca is absent, or so vestigial as to be all but invisible.

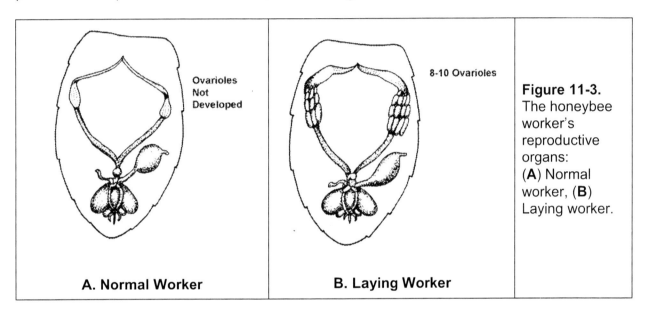

Ovarioles Not Developed

8-10 Ovarioles

Figure 11-3. The honeybee worker's reproductive organs: (**A**) Normal worker, (**B**) Laying worker.

A. Normal Worker

B. Laying Worker

The workers' ovaries are prevented from developing by activities in the hive. (More on that when pheromones are discussed in Chapter V – 16). If that pheromonal influence is lost, as through the loss or death of the queen, or by a weakening queen, the ovaries of some workers start to mature. They develop **ovarioles** and **ova**, but far fewer than the 1,000 or so ova per day in a laying queen (Figure 11-3 (B)). The workers lay their eggs but, having never mated and being incapable of receiving or storing sperm, they have only drones as progeny (Chapter VI – 22).

The **sting** of a bee (Figure 11-4) is a modified **ovipositor,** or egg-laying organ. In some parasitic wasps, like ichneumons, the egg passes down the long, slender ovipositor which is also the sting that injects poison to immobilize the prey as the egg is laid within the latter's body. The **sting** (Figure 11-4) of queen and worker honeybees is an ovipositor that has been remodeled solely for injection of poison. The egg-laying function has been lost and the sting is held out of the way of the egg as it is laid by passing to the outside via the **vagina** (Figure 11-2 (B)) to the bottom of a comb cell.

The sting of a bee is a multi-purpose organ. Workers use their stings in defence of the colony. **Queens** use their stings in offence when attacking their sister queens in the colony during the process of queen replacement following swarming. This is during the **battle royal** (Chapter V – 20). The worker's sting is straight and well barbed like the tip of a harpoon. Thus, when it penetrates skin, it cannot be withdrawn and the worker is doomed. She leaves the sting and its associated internal organs stuck in the victim when she flies off. The queen's sting is curved downward and is relatively smooth, allowing the queen to sting, withdraw, and sting again. The stings of other bees and wasps are feebly barbed or smooth (e.g., bumblebees, carpenter bees, hornets, and yellow jackets). Drones, being male, have no sting.

The **shaft** of the sting (Chapter I – 4) comprises a dorsal **stylet** and paired ventral **lancets**. The lancets in the workers carry barbs. In both queens and workers, the lancets slide back and forth against each other, supported by the stylet. The three parts enclose the canal which functions as the **venom canal** (Figure 11-4). The **poison gland** (Chapter II – 7) secretes the **venom**. (The composition of venom is discussed in Chapter X – 43.) The gland has a duct that carries the venom to the **poison (or venom) sac** (Figure 11-4) where the venom is stored. An **alkaline gland** (Figure 11-4; Chapter II – 7), also associated with the sting apparatus, produces a secretion which may lubricate parts of the sting apparatus.

Normally the sting is kept in a cavity in the tip of the abdomen. It is hinged by a complex set of small chitinous plates (the last, and highly reduced, abdominal segments (see Chapter I – 4)) connected by muscles. Coordinated contraction of the muscles (remember the last abdominal ganglion (see Chapter III – 9)) causes the apparatus of plates to rotate and swing down and out, and for the sting to be thrust backward and out. When that happens in the worker bee, the secretions of various glandular areas (Chapter II – 7) are exposed so that the stored alarm pheromone (Chapter V – 16) is released into the air and alerts other bees to the danger.

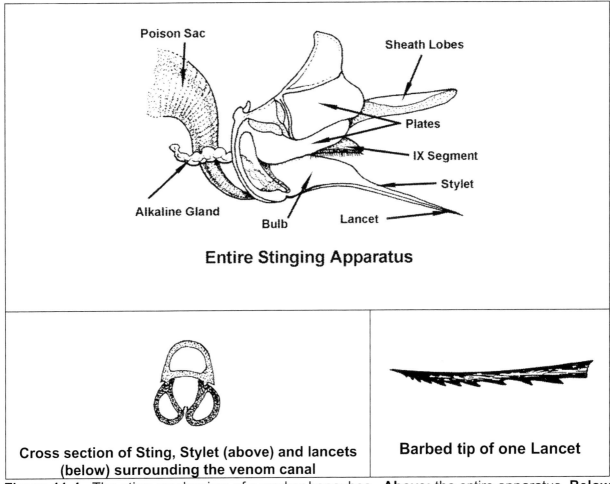

Figure 11-4. The sting mechanism of a worker honeybee. **Above:** the entire apparatus. **Below right:** a cross-section of the shaft of the sting showing two ventral lancets and the dorsal stylet surrounding the poison canal. **Below left:** details of the barbed tip of one lancet.

The Drone's Reproductive Organs

Within the abdomen of drones is a complex set of organs. **Sperm** (Figure 11-5) are made in the small, flattened, triangular and yellowish **testes**. There is a pair of these. They are at their largest, about 5 mm long, in the nearly mature pupa. The sperm are made in numerous parallel strands, and as they mature in each testis, they pass through the coiled and tubular **vas deferens** as the testes shrink. The paired **vas deferentia** enter the large, sausage-shaped **seminal vesicles**. These are storage sacs for **semen**, a mixture of sperm and secretions (Chapter II – 7) associated with the seminal vesicles and testes. The vas deferentia each are associated with the lower end of huge **mucous glands** (Chapter II – 7). From the junction of the vas deferentia and mucous glands extends the **ejaculatory duct** that enters the **intromittent** (go-between) **organ** or **penis**. Other terms for the structures involved in delivery of semen to the vagina of the queen are **aedeagus** and **endophallus**.

67

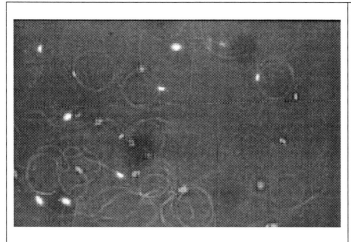

Figure 11-5. Honeybee sperm stained with fluorescent dye so that the sperm heads appear to glow. The tails can be also seen as faint threads making circles and loops (from USDA).

The drone's penis is a complex structure in the tip of the abdomen. The innermost part, to which the ejaculatory ducts lead, is the **bulb**. It extends as part of the complex intromittent organ, with various lobes and horn-like structures (Figure 11-6). During copulation, which takes place in flight, the drone mounts the back of the queen. He flips onto his back, still in flight, as the intromittent organ (penis) is everted from his abdominal cavity into the vagina (Figure 11-7) of the queen. When that happens, semen and mucus are carried out with the eversion of the penis and so placed into the queen's genital system, lateral oviducts, and vaginal cavity (Figure 11-7).

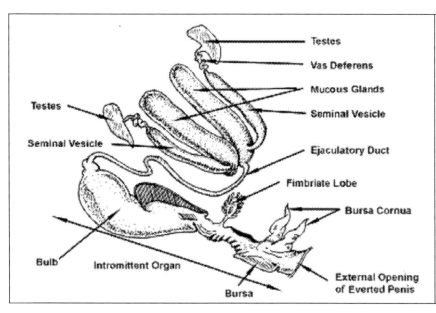

Figure 11-6. Male reproductive organs of the drone honeybee. The various appendages of the penis differ between species of bees and are used by taxonomists in identification (Chapter VI – 25). They may act in a way analogous to a complex key that fits only a certain type of lock, the female's vagina.

The everting of the penis is explosive. It is also traumatic for the drone in that during copulation he flips backwards, is dragged through the air upside down by the flying queen, and eventually falls to the ground leaving part of his genitalia behind. The drone expires. His genital remains are called **mating sign**. They are removed by the next drone to copulate or, after the last copulation, are removed by the workers once the queen is back in the hive (see Chapter VI – 21).

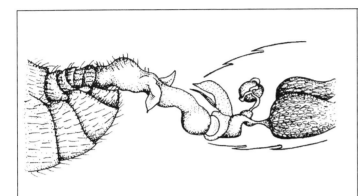

Figure 11-7. Semi-diagrammatic illustration of the arrangement of the drone's genitalia (penis or intromittent organ) (**right**) and queen's reproductive system (**left**) during mating. Note that the drone is upside down.

SECTION IV – Growing Up to Be a Bee
Preamble
From Egg to Adult, Nutrition, Hormones, & Life's Tasks

SECTION IV

Growing Up to Be a Bee
From Egg to Adult, Nutrition, Hormones, & Life's Tasks

After mating is over, sperm are stored, and eggs are being made comes the process of the development of more bees. During the course of development, **holometabolous** insects go through four main life stages: **egg, larva, pupa** and **adult**. Each looks very different from the other. Honeybees' eggs, larvae, and pupae go through their development in a highly predictable time course. They are incubated in the hive (Chapter V – 19). In the egg, an **embryo** develops. It hatches into a nondescript larva that is a feeding machine. If it is to become a queen, it dines on richer fare than do the larvae of workers or drones. The white, maggot-like larvae put on weight at a stupendous rate. Then, they stop eating, and major changes start to unfold. The larva pupates. In the motionless pupal state, huge changes take place as the larva **metamorphoses** into an adult bee. The process of **metamorphosis** is quicker for a queen, slower for a worker, and slowest for a drone. Nutrition is the key factor in female caste determination. Nevertheless, a healthy diet makes for healthy queens, workers and drones in their adult lives.

Hormones control the amazing transformations in the life of insects, including bees. There are hormones that keep larvae young, and there are hormones that signal the exoskeleton to dissolve itself and re-grow so insects can get bigger and can change shape in metamorphosis. The intricate biochemical feedback mechanisms are controlled to a great extent by neurosecretory hormones from the central nervous system (Chapter III – 9).

Once adult, worker bees have a series of tasks ahead of them. The timing is loosely controlled by a combination of the needs of the hive, the seasons, and hormones. A worker honeybee and a worker stingless bee (Chapter XI – 50) follow similar paths to their "golden days". At first they stay in the hive, attending to the needs of the larvae (brood), the queen, and the comb. Their next tasks take them from the brood chamber to attending to stores of nectar, pollen, and honey. Only after that do they venture outside the hive, first as guard bees at the entrance, and finally as foragers in the field. Most worker bees die away from home and among the flowers.

CHAPTER IV – 12

FROM EGG TO ADULT

The queen bee is the only individual who **normally** lays eggs in a colony of honeybees. However, there are some exceptions to this general rule (see Chapter V – 20).

The queen inserts her abdomen into the cells of the comb and deposits one egg neatly on the bottom of the cell. While an egg is being laid, it may be fertilized by sperm or not (Chapter VI – 22). To fertilize an egg, the queen liberates a few sperm (Chapter III – 11) from her spermatheca. The **sperm** are minute, but with vibrating tails that help propel them down the spermathecal duct and onto the passing egg. On the egg, the sperm find their way to the **micropyle**, a minute hole in the otherwise tough and protective **chorion**. At least one sperm will navigate the micropyle, then penetrate the thin **vitelline membrane** and reach the **egg nucleus**, floating in the **yolk**, to bring about fertilization. The egg is about 1.6 mm long x 0.4 mm in diameter, and is slightly banana shaped and rounded at the ends. It weighs about 0.08 mg.

Whether or not the egg nucleus is fertilized, it divides. Each daughter nucleus divides many times. The nuclei move to the periphery of the egg, under the vitelline membrane, where they become arranged in a layer of cells, the **blastoderm**. This layer encloses the yolk. The cells on the convex side of the egg become larger, and the layer starts to fold inwards as the **embryo** starts to become recognizable (Figure 12-1). The convex side is the underbelly.

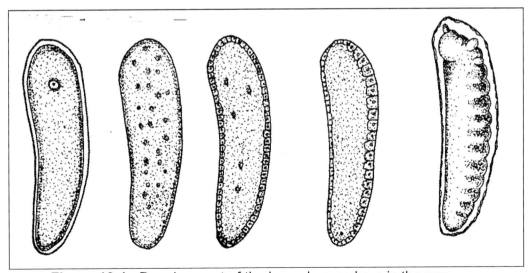

Figure 12-1. Development of the honeybee embryo in the egg.

After 3 days as an egg, the embryo shows a head, with buds for the mouthparts (labrum, mandibles, and maxillae) and the antennae. Spiracles are also apparent. Then the egg hatches and the larva emerges (Figure 12-3).

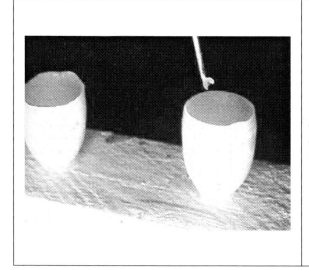

Figure 12-2. A very young larva being grafted on the point of a needle into a queen cup as the first step to queen raising (Chapter X – 44).

Over the next 6 days or so the larva grows 1,500 times in weight. It appears to have 14 segments, with a definite crease separating the dorsum from the venter. Ten segments have spiracles. Healthy larvae are pearly white.

During that period it goes through **5 larval stages, instars**, moulting as it grows from one to the next. The head has rudimentary mouthparts. Most of the interior is full of the ventriculus. The Malpighian tubules are large. The body cavity contains haemolymph and fat body. The intestine exits through the anus. The only purpose of the larva is eating and growing. It is fed and tended by the workers so does not have to fend for itself. Thus, it is an efficient and rapid food processor.

After 6 days or so, the larva occupies the entire cell and stops feeding. It spins a cocoon of silk secreted by the **silk glands**. It moults, but retains the old cuticle. The larva starts to **metamorphose** within the old larval cuticle, within the silken cocoon, within the cell, which by now has been capped with wax by attendant worker bees. This is the start of the **pupal** stage (Figure 12-4).

As metamorphosis progresses, the head becomes more structured with increasingly recognizable mouthparts, antennae, and compound eyes. The thorax develops legs and wing buds. The head, thorax, and abdomen become increasingly discrete as body parts. The abdomen develops its form, and genitalia (ovipositor and sting in queens and workers, and endophallus in drones) become recognizable.

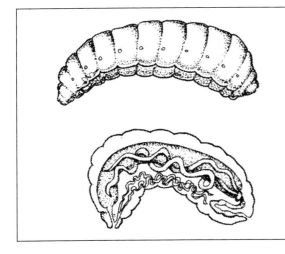

Figure 12-3. The larva. A fully-grown larva, showing the spiracles and rudimentary mouthparts. A longitudinal sectional view shows the inside, the short stomodeum (oesophagus), the very large and active ventriculus, and the short intestine (proctodeum) through which wastes pass to the anus. The metabolic wastes from large and active Malpighian tubules flow into the intestine where the ventriculus ends. The long and convoluted silk glands exit by paired pores on the rudimentary mouthparts.

After 12 days as a **pupa**, the worker bee is ready to emerge (**eclose**) as an adult at about 15 mm in length and 5 mm in diameter. The adult bee breaks out of the larval cuticle and chews its way out of its cocoon and through the capping to emerge as a **callow**. She must wait a few days for the new cuticle to tan and harden before she can fly or sting (Chapter I – 1).

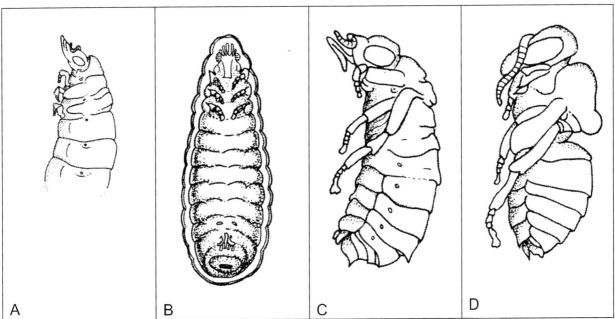

A   B   C   D

Figure 12-4. Development of the pupa. The young pupa in lateral view (**A**) showing the head, thorax and first three abdominal segments. In ventral view (**B**), the developing limbs, mouthparts, and genital region can be seen; the larval cuticle encloses the pupa. As the pupa develops, it comes to resemble more and more the adult bee (Chapter I – 1). The development of the mouthparts, antennae, compound eyes, and limbs as they increasing differentiate (**C&D**); wings as they expand from buds; the pterothorax; and genitalia can be followed through metamorphosis.

That form of insect development, from egg to larva to pupa to adult, is called **holometabolism**. Other insects, such as grasshoppers (Orthoptera) develop from eggs to nymphs that resemble the adult, to adult form without going through metamorphosis in a pupal stage (Chapter IV – 14).

Holometabolous development from egg to adult in honeybees differs in duration between castes. For European honeybees, the following schedules apply.

Egg Stage = 3 days for Queen, Workers and Drones
Larval Stage = 5.5 days for Queen
= 6 days for Workers
= 6.5 days for Drones
Pupal Stage = 7.5 days for Queen
= 12 days for Workers
= 14.5 days for Drones

TOTAL DEVELOPMENT TIME
= 16 days for Queen
= 21 days for Workers
= 24 days for Drones

To raise queens (Chapter X – 44), an important activity in modern apiculture, it is crucial to know and keep abreast of development times for the larval and pupal stages.

In other races and species of honeybees, the relative developmental rates of the different castes also hold, even though the total time differs.

At this point, it is worth a quick review of the primary differences between the castes of honeybees (Table 12-1).

Table 12-1. Characteristic Differences between Workers, the Queen, and Drones

	Honeybee Castes		
	Workers	**Queen**	**Drone**
Sex	Female	Female	Male
No. in colony	4,000–80,000	1	0–3,000
Function	Care for larvae	Lays eggs	Mate
	Produce wax	Maintains colony cohesion	
	Build comb		
	Defend colony		
	Forage for food and water		
Development Time: • *egg* • *larva* • *pupa* • *Total*	3 days 6 <u>12</u> 21 days	3 days 51/2 day <u>71/2</u> 16 days	3 days 61/2 <u>141/2</u> 24 days
Adult lifespan	+/-35 days	2–4 years	+/-35 days
Produced in Ontario in ...	February–October	May–September	May–September
Sting	Barbed, straight	Smooth, curved	Absent
Ovaries	Small: 2–12 ovarioles	Huge: 150–220 ovarioles	Absent
Spermatheca	Tiny or absent	Large	Absent
Compound Eyes	Large	Slightly smaller than in workers	Huge
Brood-food glands	Large	Vestigial	Absent
Pollen basket	Present	Absent	Absent
Pheromones	Attraction and alarm footprint	Queen substance	Mating

CHAPTER IV – 13

NUTRITION

Animals eat to live. **Nutrition** is any process whereby an organism obtains from its environment the energy and matter required for growth, maintenance, reproduction, and activities in general. Different organisms have different nutritional requirements. All animals are **heterotrophic**, meaning they rely largely on organic sources for the minerals they need: they depend on organic carbon sources for energy, and organic nitrogen and sulphur sources for proteins and body building. Honeybees are no different. They need to take in the same suite of **nutrients** as any other animal: proteins, carbohydrates, minerals, lipids (fats), vitamins and water for normal development, daily activities, body maintenance and health, and reproduction.

Honeybees are herbivorous heterotrophs. They obtain almost all their nutrition from plants, mostly from flowers, as **nectar** and **pollen** (Figure 13-1). Occasionally, honeybees may use other sources of food, such as **honeydew** secreted by plant-sucking insects such as aphids or scale insects, or by fungi. Some bees also seek out **oil** from special flowers with which they are associated also for pollination. Carnivorous bees also exist.

Pollen is chemically complex (Chapter X – 40) and most plants' pollen is highly nutritious. It contains, on average, 25% protein, 10% free amino acids, and 25% carbohydrates (including starch); the remaining percentage is made up of varying amounts of lipids, enzymes, co-enzymes, pigments, vitamins, sterols (see Chapter IV – 14), and minerals. Thus, honeybees, and, indeed most other bees and some other insects, have their entire nutritional needs met by the products of flowers. Each **pollen grain** is, in fact, a multicellular microgametophyte (contains haploid male gametes) and the mobile component of sexual reproduction in plants (Chapter X – 45).

Figure 13-1. The cells of the comb contain pollen. In one cell are visible two pollen pellets newly removed from the hind leg of a returned forager.

Honeybees, in most people's minds, are associated with honey production. **Nectar** is mostly a sugary liquid or syrup, but it contains many other constituents in minute amounts. Those may be important in overall nutrition. It is mostly nectar that is transformed into **honey** (Chapter X – 38). The chemical constituents of honey are varied and are discussed in Chapter X – 38 but, from a nutritional viewpoint, honey is mostly sugar. **Sugars** are the fuel for activity. The muscular engines of all animals require the "burning" (breakdown) of sugars to release the biochemical energy stored in the atomic bonds that hold the molecules together. We think of honey as a high-energy food, and indeed it is.

Sugars are the building blocks of **carbohydrates,** a family of organic molecules with the general formula $(CH_2O)_x$. **Monosaccharides**, such as glucose and fructose, are the simplest sugars. They cannot be hydrolyzed to a simpler structure. Disaccharides, such as sucrose (cane sugar), can be hydrolyzed (broken down) into two monosaccharides. Sugars can be assembled into larger molecules, called polysaccharides, dextrins, starch, and cellulose (the molecule that makes up the cell walls of plants). Depending on the kinds of monosaccharides and how they are linked molecularly, complex carbohydrates can be digested (hydrolyzed) in different ways. The molecules involved in digestion are called **enzymes**, and many kinds are known. In honeybees, simpler sugars can be transported across the wall of the ventriculus, into the haemolymph, for delivery to energy-consumptive tissues of the brain and muscles (Chapters II – 5 & 6).

Cellulose is difficult for many animals to digest. Herbivores, such as cattle, sheep, and termites have special microscopic, single-celled organisms that live in their guts, and these are the organisms that digest cellulose to release the energy bound in the chemical bonds. In the guts of honeybees there are also such organisms, and it is thought that they are important in the digestion of complex molecules for which the honeybee lacks enzymes to digest. Chitin is also a tough carbohydrate, but incorporates nitrogen into the individual elements of N-acetyl-glucosamine (Chapter I – 1). Sporopollenin is another tough carbohydrate that bees must tackle. It makes up the walls of pollen grains. Carbohydrates can also be bound into lipids, as glycolipids and into proteins, as glycoproteins.

During a bee's life, the bee must not only break down carbohydrates to consituent sugars and then break down those sugars to release energy, carbon dioxide (Chapter III – 10), and water, but must also build complex molecules from simple carbohydrates and the molecules that make them up. Those complex molecules, in addition to carbohydrates, are part of the exoskeleton (Chapter I – 1), internal organs, bees' wax (Chapters II – 7 & X – 39), hormones (Chapter IV – 14) and pheromones (Chapter V – 16), to mention a few.

The **proteins** used by honeybees mostly originate from pollen. Proteins are large, complex molecules (molecular weight from 10,000 to more than 1 million) that are built up from **amino acids** joined by peptide bonds. Free amino acids, and short chains of amino acids, can be found in many kinds of nectar. All proteins contain carbon, hydrogen, oxygen, and nitrogen, and most contain sulphur. Twenty amino acids are found commonly in proteins.

The proteins and amino acids ingested by honeybees become incorporated into all the body parts. The proteins must first be digested and broken down, by enzymes called proteases, to smaller constituents, such as amino acids, in the ventriculus (Chapter II – 5), and then pass, via the haemolymph (Chapter II – 6) to where they are needed for body building (e.g., of muscles, nervous tissue, other internal organs, nutritive cells for yolk in egg production, and so on).

For proper growth in honeybees, 10 amino acids are essential:
- arginine
- histidine
- lysine
- tryptophan
- phenylalanine
- methionine
- threonine
- leucine
- isoleucine
- valine

Little is known about the nutritional needs of honeybees for **lipids** or **fats**. Lipids are compounds containing carbon, hydrogen, oxygen and often other elements such as nitrogen and phosphorus. Lipids are structural components of all cell membranes and are important sources of energy. They include steroids and sterols, glycolipids, phospholipids, terpenes, fat-soluble vitamins, waxes, etc. Bees require fatty acids, sterols (cholesterol), and phospholipids in their diets for normal development. Cholesterol is essential for the production of moulting hormone during growth and development (Chapter IV – 14).

For honeybees, the dietary source of most lipids is pollen, although some nectars contain minute amounts. Bees are capable of using carbohydrates and turning them into lipids of many kinds. Nectar is used by various insects, including honeybees, as the precursor to body fat (Chapter II – 8) used as the energy source for long-distance migration, as in the monarch butterfly.

Honeybees require **vitamins** in their diets. Vitamins are micronutrients: organic substances not normally synthesized by animals. Thus, they must come from the diet. Vitamins are classified as fat-soluble (A, D, E, K) or water-soluble (B complex, C). Only a small amount of vitamins is normally required in the diet. Many **co-enzymes**, important in metabolism and interaction with enzymes, contain vitamins as part of their molecular structure. Thus, vitamin deficiency creates problems in growth, development, and especially protein utilization. The main dietary source of vitamins for bees is pollen. Pollen is generally rich in vitamins (Chapter X – 40).

Bees require B-complex vitamins (thiamine, riboflavin, pyridoxine, pantothenic acid, niacin, folic acid, and biotin) and vitamin C (ascorbic acid). Vitamins of the B-group are necessary for normal brood rearing; the hypopharyngeal glands (Chapter II – 7) are especially sensitive.

Some **minerals** are required to maintain life's functions. The most commonly found minerals in pollen are potassium, sodium, phosphorus, calcium, magnesium, copper, manganese, zinc and iron. Nectar is highly variable in its mineral constituents. Mostly, the minerals in nectar reflect the minerals found in the soils in which the plants are growing. Generally, honeybees are able to fulfill their dietary need for minerals from pollen, nectar, and the water they collect.

Water is essential to all of life's processes. It is the medium in which biochemical reactions take place, and most dietary organic and inorganic materials are soluble in water. Bees' faeces are liquid, as is their saliva. They lose water through respiration, and through evaporation. They use it for cooling themselves and their nests in hot weather. Not surprisingly, bees consume lots of water. They obtain much of their water from nectar, and they often forage for water wherever it can be found. A honeybee colony for temperate zone bees has been estimated to bring about 25 kg of water into the hive in a year. Also, in temperate climes, during winter, when the bees are restricted to the hive with only honey and pollen to eat, it must be presumed that they derive enough water metabolically from the breakdown of sugars (see above).

Larval & Queen Nutrition

What has been presented above applies to honeybee nutrition in general, and to the ultimate sources of the nutrients. Adult worker bees presumably obtain for themselves most of their dietary needs from feeding. The larvae and the queen are both fed by the workers, and both have huge requirements for nutrients (Figure 13-2). The larvae must grow, then pupate and metamorphose solely as a result of being fed. The queen's demand for nutrients is shown by her immense capacity for laying vast numbers of nutrient-rich eggs. She, too, derives her nutrition from being fed by workers. Although drones rarely feed themselves, it is presumed that they obtain most of their adult nutritional requirements, except perhaps sugars as fuel for flight, while larvae.

Brood food or **royal jelly** is produced by the hypopharyngeal and mandibular glands (Chapter II – 7) of young nurse bees (Chapter IV – 15). It has two components: a clear, highly proteinaceous liquid from the hypopharyngeal glands and a white, oily material from the mandibular gland secretions.

Royal jelly is mostly water (60–70%), and the components of the dry weight are quite variable: 17–45% protein, 18–50% sugars, 4–19% lipid, and 2–3% minerals. It contains small amounts of other valuable nutrients, including vitamins. Thus, it can be understood that it is highly nutritious, but quite variable in its constituents. Hydroxydecanoic acid (Chapter V – 16) is present, and may be the compound that prevents bacterial growth in royal jelly.

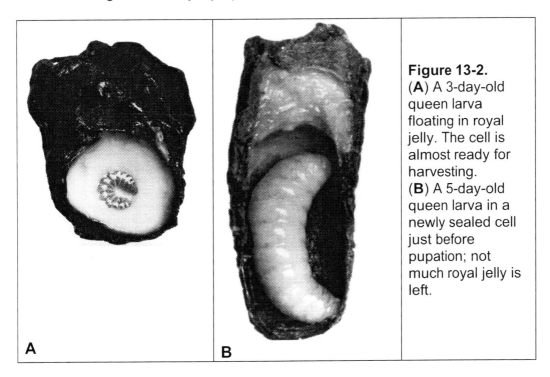

Figure 13-2.
(**A**) A 3-day-old queen larva floating in royal jelly. The cell is almost ready for harvesting.
(**B**) A 5-day-old queen larva in a newly sealed cell just before pupation; not much royal jelly is left.

All larval bees are fed royal jelly, but the duration of feeding with this rich food is important in determining the caste of the adult bee. Larvae that are destined to become queens are fed royal jelly throughout their immature lives, but larvae of workers and drones are fed royal jelly that is mixed with honey, pollen, and sometimes water. Pollen is fed directly to larval workers after the fourth and fifth day. Some investigators have suggested that there is a difference between the royal jelly fed to queens and the worker jelly fed to the other castes. The adult egg-laying queen is also fed royal jelly by her court of young worker bees.

The complete nutritional requirements and differences for all castes have not been determined. Food quality and rate of food intake are important (Table 13-1). The rate of food intake probably regulates the activity of the corpora allata (Chapters II – 7 & IV – 14) and it seems that high levels of juvenile hormone in the body of the developing larva, especially after the third day of development, induces differentiation into a queen.

Table 13-1. Different Types of Larval Food and Feeding Methods Used in Rearing the Three Castes of Honeybees

Caste	Larval Queen	Larval Worker	Larval Drone
Larval Food	Royal jelly	Royal or perhaps worker jelly, but mixed with pollen and honey	Perhaps a little different composition from royal jelly
Type of Sugar	Glucose is the major sugar	Mixture of sugars	Mixture of sugars
Feeding Methods	Heavy feeding of royal jelly	Light feeding	Receive more food than do workers

Honeybees and bumblebees are termed **progressive provisioners** because they continually provide their brood with food. Stingless bees (Chapter VI – 25 & XI – 50) are **mass provisioners** and provide their larval with all the food that is needed for growth from egg to adulthood at the time the egg is laid and sealed in a brood cell. Solitary bees are mass provisioners too. The females lay their eggs on the pollen stores that their larvae need as nutrition until they reach adulthood, as for example in alfalfa leafcutter bees and orchard bees (Figure 13-3; Chapters XI – 46, 48, & 49).

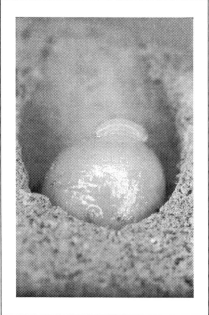

Figure 13-3. A pollen ball, with a young larva of a solitary bee on top. The surrounding material is the wall of the cell built by the parent bee.

CHAPTER IV – 14

RAGING HORMONES

Hormones are chemicals in animals and plants that regulate bodily processes such as **growth, metabolism, reproduction**, and the **functioning** of various organs. They are liberated by specialized cells in one part of the body, and have their influence in another part of the body. In humans and other animals, hormones are secreted directly into the bloodstream by **endocrine glands**—ductless glands (Chapter II – 7). The same is true of insects, but it is the haemolymph (Chapter II – 6) that carries the hormones throughout the body.

A state of **dynamic equilibrium** is maintained among the hormones. They are able to produce their effects in minute concentrations. Their distribution through the body results in responses that, although much slower than those of nervous reactions, are maintained over longer periods of time.

How do hormones work? When secreted into the body, hormones bond with specific plasma or carrier proteins that prevent them from degenerating prematurely. These carriers keep the hormones from becoming immediately absorbed by the tissues they affect (**target tissues**). Target tissues usually have receptor sites that hold the hormones until the moment they react. Then the hormone's presence initiates a biochemical cascade of reactions that influence the activity of the cells making up the tissues and organs.

In insects, including honeybees, the two most important hormones are (1) **moulting hormone (M.H.)** (**ecdysone**), mostly secreted by the prothoracic glands (Chapter II – 7) and (2) **juvenile hormone (J.H.)** (**neotinin**), secreted by the **corpora allata** (Chapter II – 7 and below). Both are involved in regulating and coordinating the growth and development of bees. **Growth** requires **moulting** because the exoskeleton does not provide the space necessary for growth. Thus, insects, and all arthropods, shed their exoskeletons and most grow by stepwise increments. In some, only the head capsule is shed before each growth spurt; the rest of the body has a flexible and expandable integument (e.g., a caterpillar). In bee larvae, larval growth also takes place in spurts, but without the shedding of the entire larval skin each time. Only when metamorphosis takes place is the larval skin entirely shed. It remains as a shell around the pupa (Chapter IV – 12). **Metamorphosis** involves all the structural changes in an organism from an egg to adult form during development. In bees, metamorphosis is the process by which the egg becomes a larva, which becomes a pupa, which eventually becomes the adult or **imago**. The processes of growth and metamorphosis require coordination and synchrony of changes throughout the entire body. Moulting and metamorphosis are controlled through the complex effects of hormonal interplay.

Metamorphosis is even more complex than growth by moulting because it not only involves moulting, but also re-organization of the body as each developmental stage is reached. The amount of bodily re-organization that insects experience varies depending on the group of insects to which it belongs. The types of metamorphosis in insects are shown in Figure 14-1.

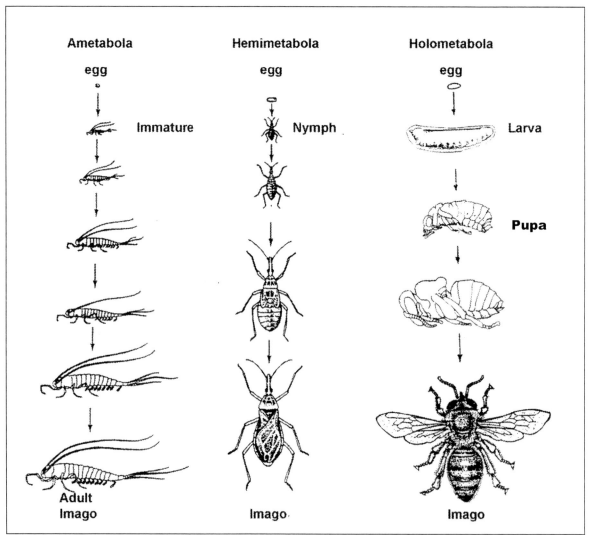

Figure 14-1. Comparison between metamorphosis in different insects: Ametabola (e.g., silver fish), Hemimetabola (e.g., true bugs), Holometabola (e.g., beetles, bees, flies, butterflies and moths). (Note that the life stages in Holometabola are not drawn to the same scale.)

Bees, flies, butterflies and moths, and beetles experience the greatest amount of bodily differentiation from egg to adult. Those are insects of the subclass **Holometabola,** which includes most insects. They pass through **complete metamorphosis**; that is, their life cycles include four stages: egg, larva, pupa, and adult, as in bees.

Other insects, the **Hemimetabola**, undergo **incomplete metamorphosis**, going from the egg to **nymph,** and then to adult. The nymphs resemble adults, but lack wings and genitalia. The wings and genitalia develop externally as the nymphs gradually increase in size, but take on a sudden burst of development from the final nymphal instar to adulthood. In the subclass **Ametabola** are primitive insects that have never evolved wings, and that change little, except in size, from instar to instar and into adulthood.

Hormonal Regulation in Insects

In honeybees and other insects, important chemical messengers are secreted by neurons and are known as **neurosecretions** (Chapters II – 7 & III – 9). The brain harbours the most important neurosecretory cells (Figures 14-2 & 14-3). The median neurosecretory cells pass their secretions, particularly **prothoracicotropic hormone**, down their axons to the **corpora cardiaca** and **corpora allata**. These are **neurohaemal organs**, so called because they provide the interface with the haemolymph by which the neurosecretions become circulated throughout the insect's body.   The secretions from the median neurosecretory cells cause a cascade of effects that control growth, development, activity, and body maintenance.

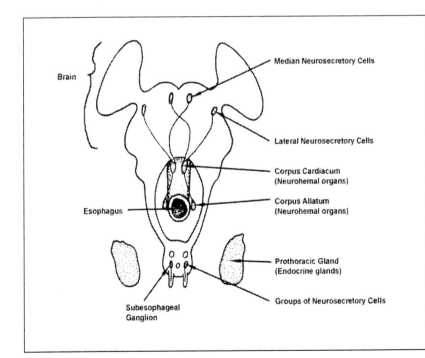

Figure 14-2. Schematic drawing of the neuroendocrine sites of hormone synthesis in the brain (cerebrum and suboesophageal ganglion), and the tracks of axons to the corpora cardiaca (dorsal aorta not shown) and corpora allata. Axons (not shown) are known to extend from the corpora allata to the suboesophageal ganglion as well.

An important consequence of the liberation of prothoracicotropic hormone is the stimulation of the corpora allata to produce another hormone, **juvenile hormone**, which is liberated into the haemolymph and influences cell division and differentiation throughout the insect body.

Also liberated from the corpora allata is prothoracicotropic hormone, but its target tissue is the **prothoracic gland** (Chapter II – 7) which becomes stimulated to produce and liberate another hormone, **ecdysone** or **moulting hormone**. As its name indicates, this hormone controls the process of moulting. It also instigates the changes that bring development of new stages, raises the metabolic rate, and increases buildup of proteins from amino acids in growing tissue.

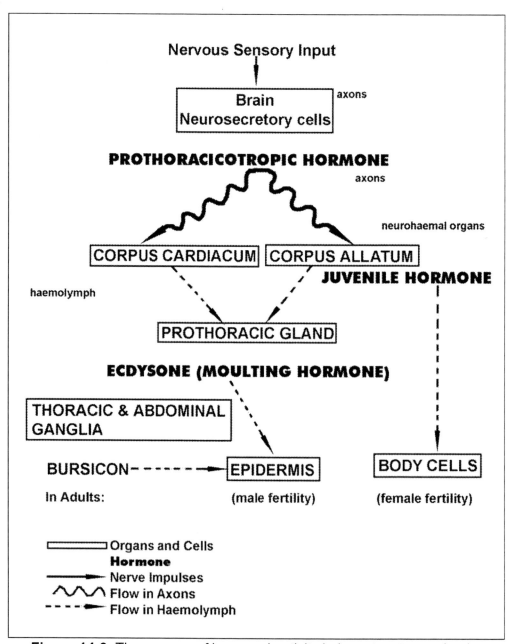

Figure 14-3. The process of hormonal activity in insect metamorphosis.

What Is the Chemistry of the Hormones Involved in Growth and Development?

Prothoracicotropic hormone is a medium-sized protein (30 kiloDaltons) and is released from the corpora cardiaca in most insects.

Juvenile hormone is quite a simple biochemical molecule (Figure 14-4). It is a terpene, similar in structure to terpenes made by plants that give some plants their characteristic odours. Terpenes are characteristically composed of short hydrocarbon (C_5H_8) chains, to which other simple chemical moieties (components) are attached. Some of the first clues about the nature of J.H. came about because Canadian forest entomologists working with spruce budworm found that caterpillars reared in Petri dishes containing brown paper substrates for the caterpillars, did not pupate. Those in Petri dishes with bleached, white filter paper did. The brown paper contained terpenes that acted like juvenile hormone.

Moulting hormone (ecdysone) is a more complex molecule, related to steroids and with almost the same structure as cholesterol (Figure 14-4). Cholesterol-like molecules in the diet, in bees primarily directly or indirectly from pollen, that are modified to make this hormone in the prothoracic gland.

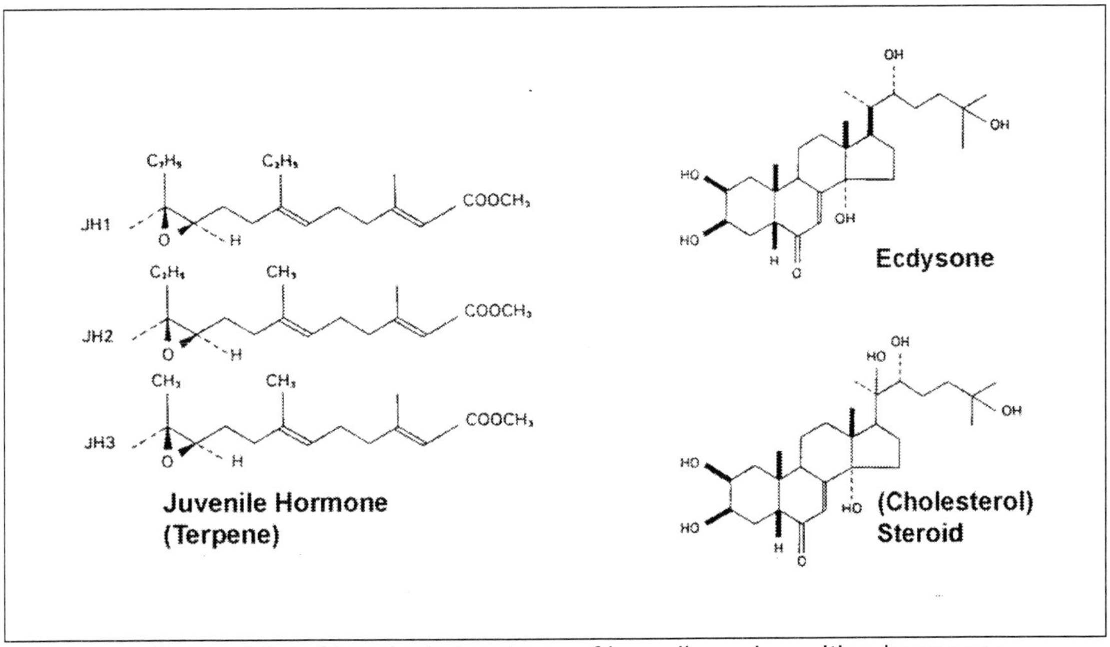

Figure 14-4. Chemical structures of juvenile and moulting hormones.

The source of another hormone, **bursicon**, is neurosecretory cells in the thoracic and abdominal ganglia. This hormone, a protein, is involved in hardening of insect cuticle. Without it, the cuticle remains soft, and wings and other appendages would not have the strength to work. It is also thought to exert some control over fertility in drones. Other hormones known to be involved in the control of moulting in insects are the peptides with names that suggest activity, **eclosion hormone** and **ecdysis triggering hormone**, but these are lesser known and have not been studied in bees.

What Is Involved in Shedding an Exoskeleton?

As noted earlier, the shedding of the exoskeleton must be coordinated with the formation of a new exoskeleton beneath the old one, and the separation of the old exoskeleton from the body and the cells lining the inside of the exoskeleton. Some of those cells are secretory (Chapter II – 7), providing oils and waxes to the outside surface. Some cells are neuronal, associated with the sensory system. Other cells are involved in the general maintenance of the exoskeleton. Thus, the shedding of the old exoskeleton requires a great deal of coordinated structural change, and physiological and biochemical activity. Figure 14-5 presents, in brief, the process of moulting that is initiated when ecdysone is liberated in sufficient amounts into the haemolymph.

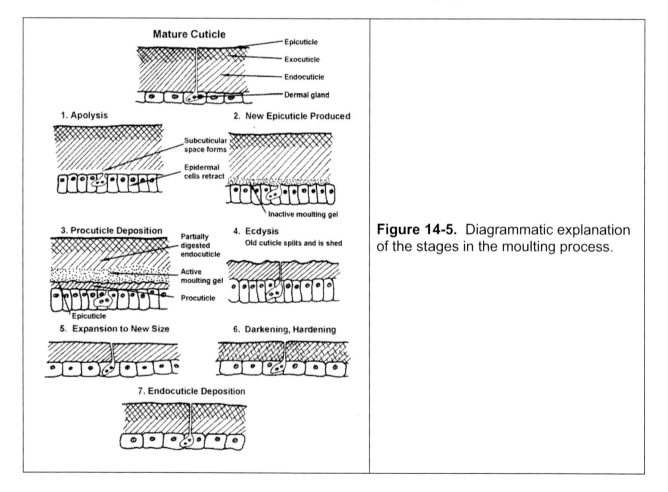

Figure 14-5. Diagrammatic explanation of the stages in the moulting process.

The Role of Hormones in Metamorphosis, Development, and Age-Related Tasks

When enough juvenile hormone is present, it dominates metamorphosis, permitting larval growth, but inhibiting pupal formation:

J.H. level high + Ecdysone: Larva (instar x) moults to larva (instar x + 1)

But when J.H. concentration wanes, ecdysone dominates, and a pupa forms:

J.H. present, but in lesser amounts + Ecdysone: Larva (last instar) moults and becomes pupa, which then completes metamorphosis, and an adult (imago) ecloses (emerges)

Juvenile hormone also controls **reproductive development.** When the concentration of both hormones decreases in insects generally, the adult activity schedule and sexual function are initiated. In bees, both queens and workers are female, and direction of development seems to be determined by the interplay of hormones and diet. Particularly important are the levels of a juvenile hormone on day 3 of larval development coupled with the quantity and quality of food fed to the larvae (royal jelly fed to larvae destined to be queen versus royal jelly and other nutrients fed to larvae destined to be workers) (Chapter IV – 13).

In worker honeybees, J.H. is also involved in the **control of age polyethism** (the schedule of tasks performed as they age (Chapter IV – 15)). Low titres of J.H. are associated with behaviours in the nest such as brood care during the first 1 to 3 weeks of the bee's adult life. Higher titres are associated with the onset of outdoor duties and of foraging (Chapter IV – 15).

Other Hormones

There are several other hormones known from insects. These are involved in various aspects of metabolism, such as fat storage and lipid release, carbohydrate metabolism, water balance, release of digestive enzymes, and other bodily functions. Even neurotransmitters, such as octopamine and acetylcholine (Chapter III – 9), can be considered hormones.

CHAPTER IV – 15

DIVIDING LIFE'S TASKS & GROWING OLD: AGING & THE DIVISION OF LABOUR

Within any social organization, individuals take on different roles. The groups of individuals with differing responsibilities may be large or small, and may segregate by age or sex, or both. Already, in honeybees, the roles of the different castes, queen, workers, and drones, have been described in general terms, but as the life of an individual bee unfolds, subtle changes in activities occur.

Ontogeny is the entire development of an organism from fertilization to completion of the life history. Bee's development from egg to adult is complex, but the behavioural changes that occur during adult life add to the overall marvel of colonial and social existence. It is during the adult's lives that the concept of the honeybee colony as a "super-organism" (Section V) starts to become apparent.

Within a honeybee colony, the **division of labour** assures that there are enough bees to perform all the tasks required for the harmonious, socially cohesive existence. Division of labour can be defined in terms of individual honeybees as the ***change in frequency at which the individual performs specific activities*** within a colony. That includes **age-regulated polyethism**, the division of labour (polyethism means many different behaviours) correlated with age and physiological development. Division of labour can also be defined in terms of the social structure of the colony as the ***segregation of duties*** according to groups of individuals that comprise the colony.

The ***segregation of duties*** in the honeybee colony is based upon (1) the sexual and physiological differences between the three castes, (2) the colony's requirements and intrinsic characteristics (genetic and physiological), and (3) external environmental conditions. The expression of the ***change in frequency at which the individual performs specific activities*** is also based on the same three criteria.

With those ideas in mind, coupled with the large numbers of individuals present in, and more or less continually being added to, a colony, one can appreciate that age-regulated polyethism (at the level of the individual) translates into **segregation of duties** by groups of bees within the colony. The concept of the colony as a "super-organism" can be expanded to colony level age-regulated polyethism (Chapters V – 20, VI – 21, & VII – 23).

The ***change in frequency at which the individual performs specific activities*** within a colony is well illustrated by **age-regulated polyethism** in individual worker honeybees. The life of the individual worker breaks down generally into three phases that reflect where duties are performed: (1) mostly in the brood area, (2) mostly in the food storage area, and (3) outside the hive. Within each of those broad categories of activities—brood and queen tending, food and stores tending, and outside jobs—there are finer divisions. Some of those finer divisions of individual age polymorphism are age related, especially at first; others are related more to the changing demands of the colony. Hormonal (Chapter IV – 14) levels, especially of juvenile hormone, play a role. Changes in the activities of exocrine glands (Chapter II – 7) occur as workers age. Those glandular activities are paralleled by expression of physical activities by the individual bees. Figure 15-1 illustrates the course of events in glandular activity and physical activity as individual honeybees divide the tasks of keeping a health colony productive.

Table 15-1. Ontogeny of Development and General Activity of Worker, Queen and Drone Castes of European Honeybees, in Days after Eclosion from Pupa

		Worker	Queen	Drone
Life Stage				
Egg		3	3	3
Larva		6	5 1/2	6 1/2
Pupa		12	7 1/2	14 1/2
Adult	**Activity:**			
		Cleans 0 - 2		
		Nurses 2 - 11	Mates 6 - 10	Mates 6 - 35
		Stores 11 - 20	Laying starts 9 - 13	
		Guards 12 - 25		
		Forages 20+		
Total lifespan		46 ± 15 days	2 - 4 years	49 ± 15 days

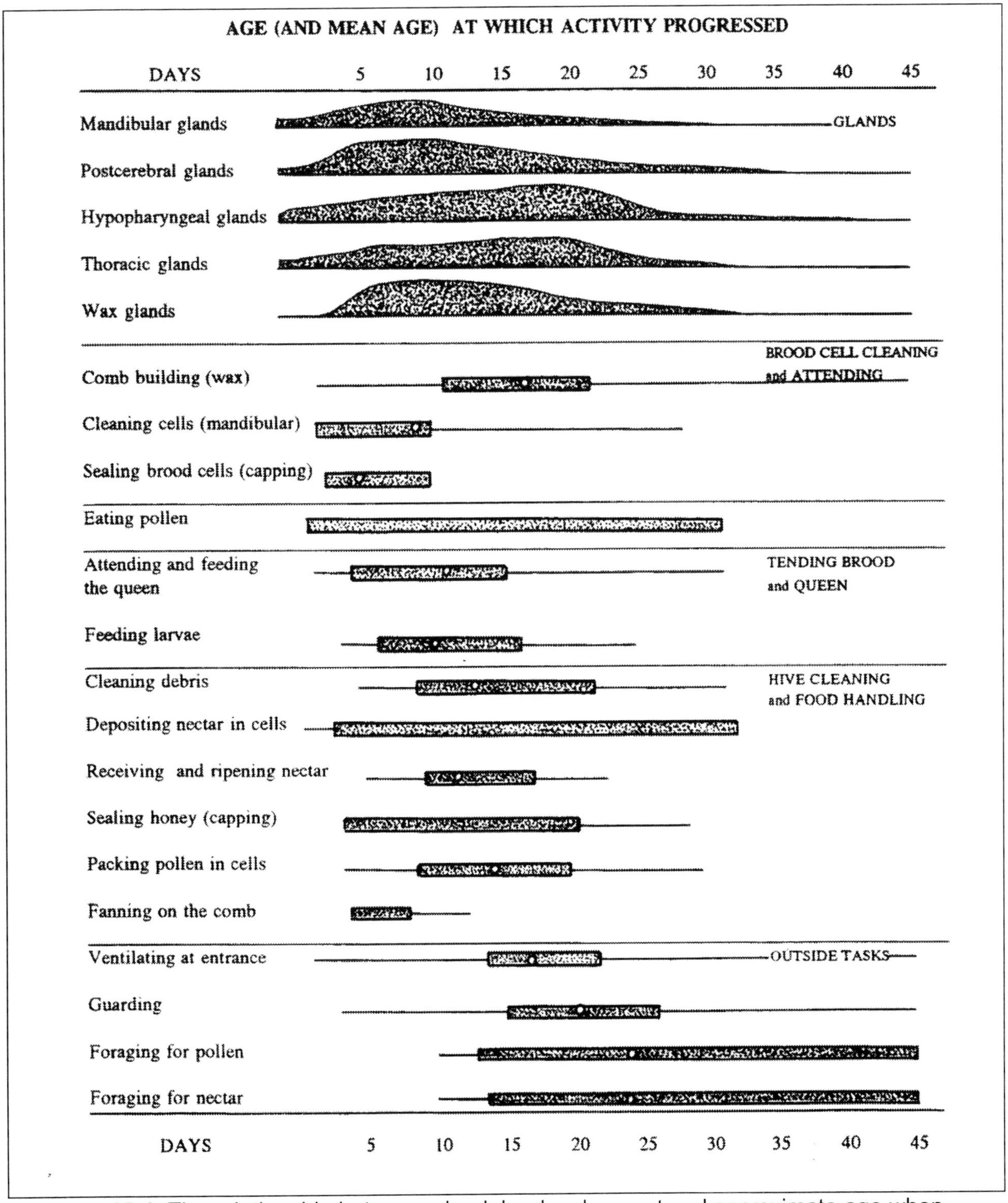

AGE (AND MEAN AGE) AT WHICH ACTIVITY PROGRESSED

Figure 15-1. The relationship between glandular development and approximate age when workers perform tasks. (Data from various sources indicate high variability in the duration and timing of some tasks.)

95

Mostly workers perform **multiple tasks** in the nest. The most common pattern for workers is that they will do a number of jobs (*generalists*) and then will tend to specialize in their respective tasks (*specialists*). Each bee is exposed to various combinations of stimuli and reacts to the most intensive stimulus. These stimuli reflect the **needs of the colony** at a particular time. The actual expression of age-related tasks is not at all rigidly programmed (Figure 15-2). Some activities, such as colony defence, are impossible for very young, callow workers (Chapters I – 1 & 4) and must be part of the potential repertoire of activities for older workers. Even so, younger, hive-bound workers take on a variety of tasks within the colony. On the other hand, if the foraging force of older workers is suddenly depleted by serious depredation or pesticide poisoning, younger workers quickly change their activities and become foragers earlier in their lives than they would normally. Age-regulated polyethism is flexible, within broad constraints.

The workers adjust their **temporal division of labour** according to the interplay of age of the worker, characteristics of worker physiology (e.g., glandular activity), spatial organization of the colony and the location of the job within or outside the nest, size of worker population, amount of brood, amount and type of stored resources, comb-building needs, weather, available forage, and colony's genetic disposition (see Chapters VI – 23 & 24).

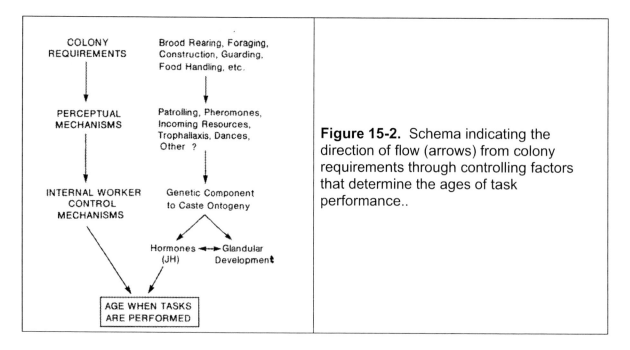

Figure 15-2. Schema indicating the direction of flow (arrows) from colony requirements through controlling factors that determine the ages of task performance..

It is interesting to consider the life of an individual worker bee in adulthood. Her first task is to break free of the larval skin and the confines of the cell in which she developed from an egg. She does that partly on her own and partly assisted by her nest-mates who may assist in removing the capping from the cell. Once the new, callow worker is free, she must wait for her cuticle to tan and harden before she can engage in physical work. That does not take long, a few hours, after which she will take up her duties. One of her first tasks is probably cleaning cells from which she and her young nest-mates have emerged. She is already in the correct location amongst nest-mates of almost exactly the same age in the brood nest. The stimuli that illicit the cleaning behaviour are probably given by the cells of the brood chamber and by her fellows. It is well known that there is a genetic component to the disposition of workers to clean within the brood chamber (Chapter VI – 23). Hygienic behaviour is important in disease control (Chapters IX – 34 & 35). Although cell-cleaning behaviour is characteristic of early days in the life of a worker, it can extend to over a month.

A young worker's next task remains in the brood chamber, as she tends to her larval nest-mates by sealing them into their cells by placing wax cappings over them. Again, the duration of this activity is highly variable. Young bees with well-developed wax glands and hypopharyngeal glands are also physiologically prepared to tend and feed the larvae. That activity may be stimulated by the odour of hungry larvae. Those activities take place mostly within the first 2 weeks of adult life, but can extend to over a month. At this same time, from their first day or so as adults, the workers are also physically in the right part of the nest and physiologically prepared to tend and feed the queen. The queen is fed mouth-to-mouth, i.e., by **trophallaxis**. It is those bees that are so important in the first steps of the dissemination of the queen's chemical signals throughout the hive (Chapter V – 16).

Generally, an adult worker bee remains in the brood nest for at least 10 days. After that, she probably becomes involved in receiving nectar from returning foragers. The bees transfer the nectar from forager to receiver also by trophallaxis. The younger worker transfers the nectar to the cells (true honeycomb) of the honey storage area in the nest (Section X) where it is cured, or ripened, through evaporation of excess water and subtle chemical changes. Once it is ripened, these workers deploy their wax glands to "cap" the honey (Chapters VIII – 32 & X – 38), potentially for long-term storage. The returning and loaded foragers solicit the attentions of the receiver workers by their behaviour, and probably also by the outside smells that they carry into the hive. Foragers may also gather pollen or plant gums, and those materials must also pass through younger receiver workers to be stored in cells in the pollen storage area, or to be used as propolis as a waterproofing and anti-microbial sealant for the hive walls (Chapter X – 42).

The adult worker honeybee's wax glands are in peak production over about a 10-day period centred on about 2 weeks old. However, those glands may remain active for much longer. Comb building is also an activity that takes the workers out of the brood nest. Again, adult workers may continue in comb-building activities for over a month.

As the adult workers age, they tend to move farther and farther from the brood nest and the food storage areas, to take on duties at the edge of the colony. One of those tasks is the cleaning of debris that has fallen from the comb onto lower parts of the nest, or onto the bottom board of a hive (Chapter VIII – 32). That activity, along with ventilating the hive, are tasks that are done by adult workers mostly in about their second and third weeks, although they may start earlier and continue much longer. Ventilating is important for maintaining the quality of the air within the hive and regulating its temperature (Chapter V – 19).

One of the activities conspicuously related to ventilating is "guard duty". At this stage in their lives, adult workers tend to be 2 to 3 weeks old. They stand at the entrance to the hive, inspecting bees attempting to gain entrance. They greet returning foragers from their own colony and allow them entry, but they attempt to repel strangers. The guard bees are able to recognize nest-mates by scent and act accordingly. Strangers may be foragers that have lost their way home to a neighbouring hive (**drifters**) or may have larceny in mind (**robbers**). If successful, robbers recruit nest-mates (Chapter V – 17) to join in the robbing-out of a neighbour's hive. Beekeepers discourage robbing by various management practices because it reduces the overall efficiency of honey production in a given apiary. Guard bees, fanning on the entrance to a hive, often expose their **Nasonov glands** (Chapters II – 7 & V – 16), releasing a plume of pheromone that their foraging nest-mates can recognize as they return. The general odour of any given hive is probably unique, so the combination of Nasonov gland pheromones and hive odour is probably a powerful familiar clue for orientation in homecoming.

It is also the older guard bees who initiate colonial defence behaviour. They are older bees, with tough, hardened exoskeletons, well-developed, sharp, hard stingers, and well-filled poison glands (Chapter II – 7). During defensive behaviour, the guard bees are joined by older workers deeper in the hive, and by foragers that are present at the time. Again, genetics plays a role in how defensive colonies may be. Africanized honeybees are notoriously prone to attack intruders to their space (Chapter VI – 24).

Up until this phase in the adult worker's life, she has not left the hive. Her first orientation flights in preparation for foraging occur when she is at least a week old, and often much older. Foraging ensues, becoming the prime activity of workers when they reach about 3 weeks old. Foraging can start as early as the end of the first week of adult life, and sometimes much later. Foraging activity also has genetic components. Africanized bees are more rigid as to when they start as foragers (at about 3 weeks). Bees have been bred to forage for nectar on particular crops (Chapter VI – 23). It is generally thought that foraging for nectar precedes foraging for other resources, such as pollen, plant gums, and water.

At this later, foraging time of the worker's life, her glandular activity has declined. During the peak of the summer, a forager's life is generally short, a matter of less than a week. She may fly from the hive a few times, or hundreds of times, in a day, depending on how rich and far away are the resources she seeks (Chapter V – 18). It seems that the total distance a forager flies in her life is about 800 km. The critical metabolic component seems to be glycogen, which is produced and stored in the flight muscles through metabolism of carbohydrates. Older foragers seem to lack enzymes in the biochemical pathway to the synthesis of glycogen.

Throughout the life of the adult worker honeybee, the level of **juvenile hormone** (Chapter IV – 14) changes. In young workers, the levels seem lower than in older workers. It is the increase in the levels of J.H. that seem to cause the decline in activity of the hypopharyngeal glands, to stimulate orientation and foraging behaviour, and to produce **alarm pheromone** (Chapter V – 16).

All in all, although there is a general pattern to the age-related activities of individual worker honeybees, the pattern is highly variable and flexible. Physiology and genetics are important determinants of intensity and timing of activity. Then the conditions within the hive and the colony's immediate requirements for resources play an important role. Catastrophes may occur. Colony reproduction, by swarming (Chapter V – 20) is also, and similarly, stimulated to different degrees by genetic, physiological, and environmental characteristics. The colony must be able to respond to the vagaries of season and weather. Opportunities cannot be taken efficiently by rigid and deterministic regimens of behaviour.

SECTION V – Coordinated Life in the Hive
Preamble
Chemical & Body Language, the Queen's Rule, Foraging, & Keeping Comfortable

SECTION V

Coordinated Life in the Hive
Chemical & Body Language, the Queen's Rule, Foraging, & Keeping Comfortable

Although honeybees and stingless bees (Chapter XI – 50) exist as individuals, they cannot fulfill their lives outside the colony. In that way they are different from most animals; even most bees live as individual, mostly solitary, insects—nesting aggregations and mating notwithstanding. The colony of eusocial insects (Chapter I – 1) is a sort of "super-organism". The individual members of the colony can be likened to cells of a body, even their own bodies. How do they maintain integrated and cohesive activity?

The queen rules by chemical force. Her pheromones influence every facet of colonial activity. She inhibits the colony from reproduction by swarming (Chapter V – 20). Her influence is involved in the complex interactions within the colony that keep the workers from developing ovaries. Pheromones influence genetic expression and behaviour. Her scent is attractive to drones at mating time and to workers during swarming. The workers produce their own suite of chemical signals that help integrate their activities through nest-mate recognition, in aging, foraging, and colony defence. Even the brood communicates with the colony by odours.

The value of **body language** is shown superbly by **dance communication** in honeybees and stingless bees. Through dance and song, they inform each other of the distances and directions to resources. Once they are in the field foraging, though, workers act more as individuals. They learn their local landscape by vision and smell to aid in navigation to and from areas of forage. Once in the field, they forage efficiently along routes they have come to know and on flowers that they have learned to manipulate. On flowering arrays of herbs, shrubs, and trees, they are systematic in expending as little energy as they can to harvest as much as they can. That is **optimal foraging**!

Within the colony, bees are frugal in their use of energy (honey) and other food. European races of the Western honeybee (*Apis mellifera*) remain active all winter. When the queen is not laying, the colony is cooler than when she lays and brood is to be incubated. The colony circulates air to maintain a healthy environment in winter and summer. The colony is its own biological central heating and air conditioning unit. Some bees (*A. cerana*) use heat to kill alien invaders to their hives, others (*A. dorsata*), with colonial coordination dump faeces together to ease thermoregulatory cooling when it is exceptionally hot.

Section V – Coordinated Life in the Hive
Chapter 16
Chemical Communication, Pheromones, & the Queen's Rule

CHAPTER V – 16

CHEMICAL COMMUNICATION, PHEROMONES, & THE QUEEN'S RULE

Chemical signals are an important part of communication between individuals of the same species, and between individuals of different species. The chemicals involved are produced in glands that secrete to the exterior (exocrine glands, see Figure 16-1; Chapter II – 7); they are called **semiochemicals** ("signal chemicals" involved in behavioural and physiological interactions between organisms) or **infochemicals**, to be more specific. The term semiochemicals can be broadly defined to include toxins. Human beings give off and receive infochemicals.

Infochemicals fall into several broad categories, named according to how they benefit the emitter, receiver, or both. **Kairomones** benefit the receiver, as for example, the particular smell of a plant allows a plant-specific herbivore, such as a monarch butterfly, to distinguish its host plant (milkweed) from other plants. Kairomones may seem disadvantageous to the emitter, but, then again, many generalist herbivores avoid milkweed (it is poisonous to many animals), so the overall depredation by herbivory may be lessened. **Allomones** benefit the emitter, as for example a chemical signal that acts to repel a predator. The spray of a skunk is an obvious example, and the chemicals exuded by milkweed act as allomones in repelling many generalist herbivores. **Synomones** benefit both the emitter and receiver. In the life of bees, floral scents act as attractants to rewards (nectar or pollen, or both), and the bees act as pollinators and so benefit the plant (Chapter X – 45). Those terms for **allelochemicals**, chemicals that provide information between species, must be used in the context of the emitter and receiver, as illustrated by the dual function of milkweed chemicals, which expands to three-way function when toxicity is included.

Infochemicals used within species (intraspecifically) are termed **pheromones**. They stimulate a response in animals of the same species. In many animals, especially insects, **intraspecific behaviours**, such as aggregation, sexual attraction, dispersion, defence, etc., are regulated by chemical stimuli. For honeybees, pheromones constitute the chemical language of **intraspecific communication** in the colony. It is estimated that honeybees produce at least 36 pheromones. In some insects, including honeybees, pheromones may also have allelochemical activity. The scent of an active honeybee colony acts as a kairomone in attracting pests, such as wax moths and hive beetles (Chapter IX – 35), and predators, such as bears and skunks (Chapter IX – 36).

Honeybees have evolved a complex **chemosocial** lifestyle by which most of the information and messages involved in the daily life and behavioural cohesion in the colony are generated by pheromones (see Table 16-1). They are emitted by special

Section V – Coordinated Life in the Hive
Chapter 16
Chemical Communication, Pheromones, & the Queen's Rule

glands and generally over the body surface (Chapter II – 7). They are received by chemoreceptors and detected as odour (olfaction) or taste (gustation) (Chapter VII – 27). They are difficult to study because they are produced in minute amounts, but with huge strides in analytical chemistry and instrumentation in recent years, more and more is being learned of the diversity of pheromones and their chemical structures. If the pheromone can be artificially synthesized, or obtained in large enough amounts from natural sources, they can be tested by **bioassay** to determine their role in eliciting specific behaviour. Test bees are exposed to the chemical and their reactions noted.

There are two basic types of pheromones, releaser and primer. **Releaser pheromones** trigger almost an immediate response by the receiving individuals. They act via the central nervous system to produce a quick behavioural response. Most pheromones fall into that category. **Primer pheromones** "prime" the receiver individual to exhibit an altered behavioural activity at some future time (e.g., queen pheromones). They act to modify the physiological state of the animal.

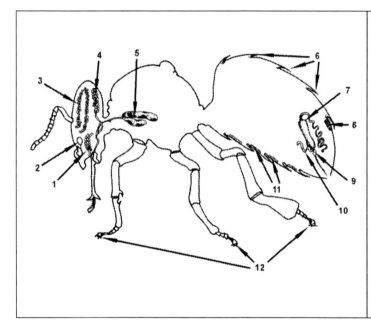

Figure 16-1. Diagram of the exocrine glands in a composite queen–worker bee (Chapter II – 7). (1) Postgenal (= hypostomal gland), (2) Mandibular glands, (3) Hypopharyngeal (= maxillary) gland, (4) Cephalic labial (= salivary) gland, (5) Thoracic labial gland, (6) Tergal glands, (7) Poison gland, (8) Nasonov gland, (9) Koschevnikov gland, (10) Dufour's (alkaline) gland, (11) Wax glands, (12) Tarsal (Arnhart) gland. Glandular areas associated with the complex of abdominal segments associated with the sting apparatus are not shown (Chapters I – 4 & II – 7).

The **worker-produced releaser, recognition pheromones** of the Nasonov gland (Figure 16-2) are the best known orientation clues. Like most pheromones, Nasonov secretion is a blend of chemicals, in this case seven. Although (*E*)-citral and geranic acids produce the greatest responses in bioassay by field tests and electroantennogram (Section VII), geraniol is produced in the greatest amount. However, the greatest response results from exposure of the bees to the full blend. The importance of Nasonov secretions is in nest and nest-mate recognition at the entrance to the hive, and in orientation and clustering during the process of swarming (Chapter V – 20).

Section V – Coordinated Life in the Hive
Chapter 16
Chemical Communication, Pheromones, & the Queen's Rule

Table 16-1. Pheromones Produced by Adult and Larval Honeybees, Caste and Life Stage (Including Comb), the Glands Involved, the Chemical Names (*general type in italics*; **abbreviations in bold**), and Their Effects or Functions

Pheromone	Gland	Chemicals	Functions
Worker-produced: Nasonov	Nasonov	*Terpenoids:* • Geraniol • Nerolic acid • Geranic acid • (E)-citral • (Z)-citral • (E-E)-farnesol	Orientation Colony and nest recognition
Footprint	Arnhart	?	Orientation
Alarm	Mandibular	• 2-Heptanone	Alarm and defence
Alarm	Sting and sting apparatus	• Isoamyl acetate • 2-nonanol • *N*-butyl acetate • *N*-hexyl acetate • Benzyl acetate • Isopentyl alcohol • *N*-octyl acetate • Decenyl acetate • (Z)-11-eicosen-1-ol	Alarm and defence (blend, and proportions of chemicals vary between species)
Dufour	Dufour's or "alkaline"	Alkanes with odd numbers of carbon atoms	Recognition, Egg recognition
Rectal gland	Rectal gland	?	?
Forage marking	? Arnhart	?	Orientation at food sources
Recognition	?	?	Kin and/or colony recognition

Nasonov pheromone:

CH₃— CH₃
   C=CH-(CH₂)₂-C=CH₂OH
CH₃—
geraniol *trans* form (*primary terpenic alcohol*)

CH₃— CH₃
   C=CH-(CH₂)₂-C=COOH
CH₃—   H
geranic acid *trans* form (*terpenic acid*)

CH₃— CH₃
   C=CH-(CH₂)₂-C=C-CHO
CH₃—   H
citral or geranial *trans* form (*terpenic aldehyde*)

Alarm pheromone - sting:

CH₃—
   CH-(CH₂)₂-O-CO-CH₃
CH₃—
isoamyl acetate or isopentyl acetate (*alcohol acetate*)

Alarm pheromone - mandibular:

CH₃(CH₂)₄-CO-CH₃
heptan-2-one (*aliphatic ketone*)

Section V – Coordinated Life in the Hive
Chapter 16
Chemical Communication, Pheromones, & the Queen's Rule

Source	Gland	Compounds	Functions
Queen-produced: Queen substances, 25 or more pheromones	Mandibular	• 9-keto-(E)-2-decenoic acid (or 9-oxodecenoic acid) (**9-ODA**) • 2 isomers of 9-hydroxy-(E)-2-decenoic acid (**9-HDA**) • methyl p-hydroxybenzoate (**HOB**) • 4-hydroxy-3-methoxyphenylethanol-diacetate (**HVA**)	- Induces retinue behaviour in workers - Swarm cluster stabilization - Drone attraction - Worker attraction to swarms - Swarm cluster stabilization - Stimulates release of Nasonov pheromone - Stimulates worker foraging
	Other glands: MO throughout the body; CA concentrated in the head; PA concentrated in abdomen; LEA concentrated in thorax and abdomen	• methyl (Z)-octadec-9-enoate (methyl oleate, **MO**) • (E)-3-(4-hydroxy-3-methoxyphenyl)-prop-2-en-1-ol (coniferyl alcohol, **CA**) • hexadecan-1-ol (**PA**) • ($Z9,Z12,Z15$)-octadeca-9,12,15-trienoic acid (linolenic acid, **LEA**)	
Dufour	Dufour's or "alkaline"	• Alkanes with odd numbers of carbon atoms • Long-chain esters	Recognition, Egg recognition
Olfactory	Koschevnikov	?	Worker attraction
Tergite	Tergite	?	- Stabilizing "retinue" - Drone attraction and copulation - Inhibition of worker ovary development - Inhibition of queen rearing
Immature	?	?	- Inhibition of queen rearing
Footprint	Arnhart	? at least 12 compounds	- Inhibition of queen cup construction
Drone-produced: Marking	Mandibular	?	- Marking of congregation areas - Attracting drone to drone

Section V – Coordinated Life in the Hive
Chapter 16
Chemical Communication, Pheromones, & the Queen's Rule

Brood-produced: Brood	?	Glyceryl-1,2-dioleate-3-palmitate (blend of 10 fatty-acid esters)	- Stimulate foraging for pollen - Brood recognition (incubation pheromone from pupae) - Inhibition of worker ovary development
In young diploid drone larvae: "Cannibalism pheromone"	?	?	- Attracting workers to eat the larvae
Comb-produced: Hoarding	N/A	?	Increase nectar storage

Figure 16-2. A worker honeybee fanning with its Nasonov gland conspicuously presented.

Other orientation pheromones, such as footprint pheromone, are less well understood. It has been suggested that footprint pheromone left by bumblebees on flowers as they forage has two functions: at first it marks the flower as having been visited recently and seems to be repellent, possibly indicating that the rewards have been depleted. As the pheromone is exposed to the air, it changes and later becomes attractive, possibly indicating that the flower was previously rewarding and may have replenished the reward.

Section V – Coordinated Life in the Hive
Chapter 16
Chemical Communication, Pheromones, & the Queen's Rule

Workers produce a complex blend of chemicals associated with colony defence and stinging. The mandibular glands and various glands of the sting and sting apparatus lend their chemical products to eliciting a suite of activities. The chemicals cause workers to become agitated, act defensively, and to orient to the location of the pheromones. If the workers are presented with just the chemicals, they generally do not sting unless the object is moving. As with Nasonov secretions, the individual chemicals may elicit weak to moderate responses, but it is the full blend that results in the greatest activity. Some components, such as 2-heptanone, may have their primary role in repelling robber bees and perhaps other threats at the hive entrance. Isoamyl acetate, also known as banana oil, is known to strengthen stinging and general defensive behaviour. The smell of bananas around an agitated colony of bees is a signal to the beekeeper to retreat!

The reason that **alarm pheromone** is such a complex chemical blend is not fully understood. Perhaps some of the compounds, such as 2-heptanone, also have other functions as yet undiscovered. Workers of different ages produce the different chemicals at different ages. Isoamyl acetate is produced mostly by 2- to 3-week-old workers, 2-nonanol is found only in older workers and then not in all. That chemical may be involved in foraging behaviour. Very young workers do not respond much to alarm pheromone, or to queen odours. They respond strongly once they are about a week old. The blends of alarm pheromone differ between the species of *Apis* that have been investigated: Decenyl acetate is present in *A. dorsata* and *A. cerana*; octyl acetate is present in *A. florea* as well as *A. dorsata* and *A. cerana*.

The main pheromone produced by the queen can be called **queen substance**. It is produced in many glands of the body, some of which have not been identified. **Queen mandibular pheromone (QMP)** is part of the queen substance and is a blend of five chemicals. The full effects of the presence of a healthy queen in a colony cannot be elicited by QMP, nor even by a mixture of all the chemicals so-far known from queen substance. All the chemicals in QMP are effective to some extent individually, but some of the other chemicals have no effect unless combined with QMP.

The first demonstrated function of 9-ODA and 9-HDA was in the suppression of queen rearing in a colony. Queens were removed from experimental hives in the apiary at Simon Fraser University. Some were left queenless, but others were given one queen equivalent of QMP. In the queenless colonies, the workers started making queen cells within 2 days, and after about 6 days, those colonies had on average eight queen cells. Adding the QMP to the other queenless colonies suppressed their building queen cells for about 4 days, but after that they started. By the end of the experiment, both sets of experimental colonies had about six queen cells on average. The effectiveness of the QMP lasted for several days, even though it is known to dissipate quickly in active colonies (e.g., it has been suggested that half of the queen's QMP

Section V – Coordinated Life in the Hive
Chapter 16
Chemical Communication, Pheromones, & the Queen's Rule

dissipates within about 20 minutes). Queen **footprint pheromone** may also be involved.

It was generally believed that 9-ODA and QMP also acted as a primer pheromone in inhibiting the development of workers' ovaries. Results of experiments in which artificial QMP was introduced into colonies have shown that the issue is much more complex. Even as much as 10 queen equivalents of QMP placed into queenless colonies did not suppress ovarian development. It is likely that complete queen substance is required, plus perhaps the presence of pheromones from the brood. In any case, it is presumed that the effect on the workers comes about through inhibition of the **corpora allata** and their hormones (Chapter IV – 14). It is known that QMP affects behavioural gene expression, activating "nursing" and repressing "foraging" genes.

It is no surprise that the queen's pheromones are important during mating; both 9-ODA and 9-HDA are attractive to drones. The former can attract drones from as far away as 60 m; the latter, along with **tergal gland chemicals**, seems to release copulatory behaviour.

During swarming, the presence of the queen is essential. The same chemicals are involved in cohesion of the swarm during flight, and its stabilization when settled (Chapter V – 20). Workers in swarms react to 9-HDA by exposing their Nasonov glands, further adding to social integration.

Within the hive, the queen is highly attractive to workers, who tend, groom, and feed her. QMP is very much involved in that crucial repertoire of behaviours and, no doubt, so are the other secretions that make up Queen Substance. The queen produces a synergistic, multi-glandular pheromone blend of at least nine compounds for retinue attraction. That is the most complex pheromone blend known for inducing a single behavioural outcome in any organism. It is via the queen's retinue that her chemical messages are distributed throughout the hive.

How do the chemicals diffuse through the colony? To answer that question, a series of experiments with **radiolabeled 9-ODA** as part of the synthetic pheromone blend were conducted, again at Simon Fraser University. Figure 16-3 shows the results of the experiment. The **messenger** is a worker which receives pheromone directly from a queen. The worker may touch the queen, and so become a carrier of pheromone by **antennating** or **licking**. Although it appears that little pheromone becomes transferred from the queen to the messenger worker, one must remember that the queen's retinue is continually changing, and that the bees of the retinue move throughout the brood chamber. Even the bees not involved in activities directly associated with the brood pass through the brood chamber to deliver their loads (foragers), to transport the resources gathered to storage above the brood chamber, or to assume guard duty at the hive's entrance.

Section V – Coordinated Life in the Hive
Chapter 16
Chemical Communication, Pheromones, & the Queen's Rule

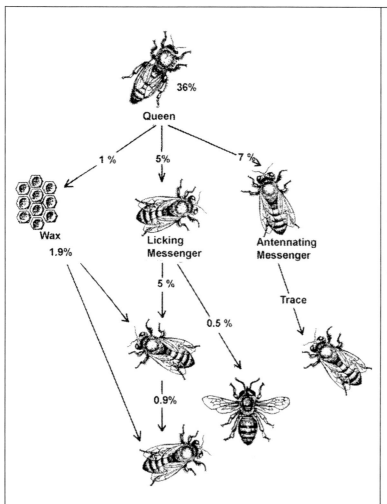

Figure 16-3. Model of pheromonal transmission within a colony. The numbers give the percentage of the queen's daily production that is transmitted in the indicated direction.

Dufour's gland and its secretion have been somewhat of a mystery. The gland secretes its alkaline products into the vaginal cavity, and it has been assumed to be deposited on the eggs as they are laid. Indeed, Dufour's secretions allow worker bees to distinguish between eggs laid by the queen, which are attractive, and those laid by workers. The complex of as many as 24 chemicals differs between workers in queen-right colonies and workers of queenless colonies. In the latter, the workers' Dufour secretions are similar to those of a healthy queen.

The **pheromones of drones** are not well studied. They seem to be produced early in a drone's adult life and are stored in the head. They may be released and attract drones or queens, or both, to congregation areas for mating (Chapter VI – 21).

Section V – Coordinated Life in the Hive
Chapter 16
Chemical Communication, Pheromones, & the Queen's Rule

Brood odours seem to have several functions: they may stimulate nurse workers to provide food; may stimulate foraging in older workers; and they may be involved in the suppression of ovarian development in workers. The pheromone is a 10-component blend of fatty-acid esters. Experimental addition of brood pheromone to colonies increases the proportion of foragers that collect pollen rather than nectar. For suppression of ovarian development, the presence of a healthy queen is also required. Only if the colony has been queenless for a few days do laying workers make a last attempt to pass the colonies genes to another generation of bees through production of drones (Chapter VI – 22).

Kin selection is important in honeybee colonial cohesiveness. Recognition pheromones, such as emitted from the Nasonov gland, function to allow relatives to orient homeward to relatives. The returning forager nest-mates are recognized by the guard bees and allowed entry to the hive. When beekeepers requeen their colonies (Chapter VIII – 32), they are careful to allow a prolonged period of introduction because if they do not, the resident workers detect the strange queen and usually kill her. At a more subtle level, it has been suggested that honeybees are able to distinguish full-sisters from half-sisters (the latter originate from eggs fertilized by sperm from different drones), but the evidence for such discrimination is weak.

Some Practical Considerations

QMP can be made synthetically to include the five chemicals in the correct proportions. It has been used to keep packaged bees calm during shipment, especially if the package is queenless. It has also been advocated as a means to attracting foraging workers to pollinate various crops (Chapter X – 45), including apples, pears, cranberries, blueberries, kiwifruit and onion. Although the results of some experiments suggest benefit with more bees foraging longer at flowers followed by greater crop production, most are generally inconclusive. In some trails, the experimental design was flawed by lack of proper controls and pseudoreplication. This means that application of a pollinator attractant to parts of orchards or crop fields would be expected to detract from pollination on untreated parts for the same area; thus the control (untreated) areas are not properly independent of the treated areas, which, in turn are not independent of each other (pseudoreplication).

Synthetic **Nasonov pheromone** can be used to trap swarms. This can be useful in areas where Africanized bees are problematic because of their swarming, migrating, and highly defensive behaviours.

CHAPTER V – 17

DANCE COMMUNICATION

In the broadest sense, **communication** is an action by one animal that influences another animal. More specifically, it can be described as any process of transmitting and receiving information and messages.

Communication occurs via many media: visual, auditory, tactile, and chemical. **Visual** messages may be communicated by colour, posture or shape, movement and its timing. **Sound** communication is limited (it may surprise you to know), for the most part, to arthropods and vertebrates. Also, behaviour of many animals is based on **chemical communication**. In this case, transfer of information occurs by **infochemicals** (Chapter V – 16) that are produced by one animal and influence the behaviour of another animal of the same or different species.

Non-chemical communication between people is an outgrowth of behaviours that have developed over millennia of expression. Some theories hold that language developed as a consequence of group activities, such as working together or other social interaction. Other theorists believe that language developed from basic sounds that accompanied gestures.

Multiple mechanisms of communication have evolved in honeybees and stingless bees (Meliponini) (Chapter VI – 25 & XI – 50). The different stimuli (chemical [Chapter V – 16], sound, touch) have various uses in colony functions. Effective communication involves the activities of the "receiving" bees as much as of the "sending" bees. The non-chemical communication in eusocial bees (Chapter X) has been referred to as a language, the **dance language**. However, some ethologists and psychologists prefer to restrict the word "language" to communication involving novel and abstract information, including the association of ideas.

Dance Communication "Language" of Honey Bees

The naturalists of ancient Greece and Rome understood that somehow honeybees were able to recruit fellow foragers to a source of sweet liquid. Can we assume that Aristotle (384–322 BC), while eating something sweet, was visited by a honeybee that ate and flew away, only to be replaced by several honeybees within a few minutes? That suggested to him the honeybees were conveying information about the resource. Once glass observation hives were widely used, more observations could be made. The Reverend Ernst Spitzner in Germany in 1788 recorded that returning foragers twirled as if with joy upon returning to the colony.

Two Nobel laureates have been instrumental in the elucidation of what the method of communication is, and what the twirling meant. **Maurice Maeterlinck** (1862–1949; photo on right) was awarded the  Nobel prize for literature in 1911. His most popular book was "*Vie des Abeilles*" (Life of Bees). **Karl**  **Ritter von Frisch** (1886–1982; photo on left) shared the Nobel prize in medicine in 1973 with two other eminent ethologists, Konrad Lorenz and Nikolaas Tinbergen. Von Frisch published the results of his first research into the dance language in 1923. His highly influential book "*Tanzsprache und Orientierung der Bienen*" was published in 1965 and in 1967 translated as "*The Dance Language and Orientation of Bees*".

Worker bees communicate information about the location of food sources through a series of body movements called **dances** that are performed on the comb. Von Frisch described three types of bee dances in European honeybees that indicate distances to food sources. Those dances apply to all species and races of honeybees, i.e., the genus *Apis*, although variations in the themes are known. They depend on the species and race of honeybee. There are several other dances known to have roles in communication in the colony, but not for location of food. The food location dances fall into three categories: **round dance** (Figure 17-1 (A)), **transitional dance,** and **waggle dance** (Figure 17-1 (B)). The waggle dance is so named because the bees waggle their abdomens from side to side when making the run between the two loops of the dance.

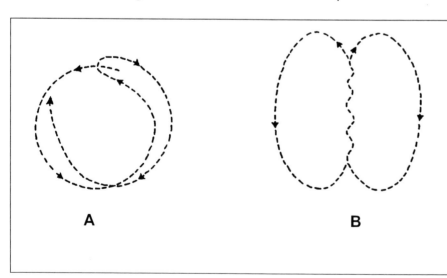

A   B

Figure 17-1. (A) The **round dance** and **(B)** the **waggle dance.** The dancing bee follows the path indicated by the arrows. The diameter of the dance is about 20 mm, so one can appreciate that the dancing bee is turning sharply as she changes direction. In the round dance, the bee reverses direction every circuit or two.

Honeybees perform the **round dance** when the source of food is close to the hive. She runs around in narrow circles on the comb, changing direction after one or two circuits. She may dance for several seconds or up to about a minute in one spot before moving to another location on the comb and dancing there. She can indicate the quality of the food source by the vigour or liveliness of the dance, and by the odours she may carry in from the food source (e.g., the smell of flowers). During the dance she is followed and antennated by nest-mates who are thus recruited to go out and forage close to the hive, and to find the food source through the scent it emits.

The **transitional dance** is sometimes referred to as the **sickle** or **crescent dance** (Figure 17-2).

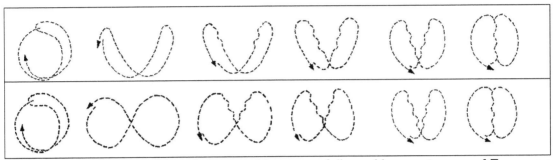

Figure 17-2. The **transitional dance: Above:** as followed by most races of European honeybees (the sickle shape is more evident); **Below:** as followed by the Carniolan race (*Apis mellifera carnica*) (in which a figure eight is characteristic). The dancing bee follows the path indicated by the arrows. The diameter of the dance is about 20 mm, so one can appreciate that the dancing bee is turning sharply as she changes direction.

In the transitional dances, the pattern changes from the round to the waggle dance. The transitional dances become evident when foragers of European races return from foraging at a source at least 6 m away (Table 17-1). Although the source of food is nearby, especially at the initiation of the transitional dance, the dance does indicate the direction to the source. Direction to the source is indicated by the imaginary line that bisects the two arms of the dance, whether together they make a crescent shape or a figure eight.

Table 17-1. Some Generalized Relationships between the Type of Dance Used in Food Location by European Races of the Western Honeybee (*Apis mellifera*), and the Distance to the Food Source in Metres

Round Dance	Distance to Food source
Apis mellifera ligustica	< 8 m away
Apis mellifera caucasica	< 14 m away
Apis mellifera carnica	< 16 m away
Transitional Dance	*ca.* 10–100 m away
Waggle Dance	> 90 m away

In the **waggle dance**, honeybees communicate information about the **distance, direction**, and **quality of food resources**. The waggle dance is used once the source of forage is more than about 80–100 m away (Table 17-1). One may readily appreciate that a honeybee in search of a source of forage at a distance of several kilometres would be hard pressed to find that source unless she had some inkling of how far to go and in which direction. The waggle dance gives information about the distance by the **dance tempo**, often measured as the number of straight runs per 15 seconds. The **tempo** is correlated with the **distance** to the source of forage (Figure 17-3). The waggle dance also gives information about **direction** by the **orientation of the straight run** (i.e., the line bisecting the symmetrical pattern of the dance; see below).

A dancing bee, on returning from a foraging trip, moves in a narrow semicircle to one side, then turns sharply and moves in a straight line, while waggling her abdomen from side to side, along the imaginary mid-line of the dance pattern. She then makes a semicircle in the opposite direction completing the full pattern. She runs again in a straight line until again she reaches the starting point. While running the straight portion of the dance, she waggles her abdomen rapidly (it is just visible, rather than a blur, to the naked human eye (13–15 times/sec)). After a few dance circuits, she stops and distributes food to her nest-mates following her movements. Sound communication is important in that phase of communication by the dance (see below).

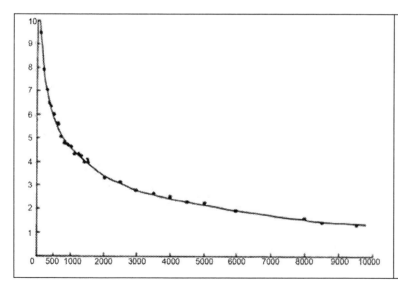

Figure 17-3. Dance tempo and distance. The relation of the tempo of the waggle dance as measured by the average number of circuits per 15 seconds and the distance (metres as given on the horizontal axis) to the source of forage.

The dance tempo changes rapidly to indicate distances between about 100 m and 2 km. After that, the rate of change in the tempo with distance becomes less pronounced. Thus, one can appreciate that at fairly close distances of less than 2 km, the tempo of the dances made by the returning forager provides accurate and precise information to the recruits using that information to embark on their own foraging trips. Then, although it is important for central place foragers (i.e., those who forage out from a home base to which they must return; Chapter V – 18) to know the distance to forage that is out of sight and out of range for scent, direction also matters. Flying away from home, and travelling several kilometres in the wrong direction, in expectation of being able to forage would be a risky business, possibly with a high likelihood of failure. The greater the distance, the more energy must be used in the round trip, and the more important directionality is.

Understandably, as the dance tempo slows, the number of times a bee waggles her abdomen during the straight run part of the dance increases. There is a more or less linear relation from about 10 full waggles of her abdomen at 500 metres to about 50 full waggles at about 4.5 kilometres. Thus, there is a second interpretable component to the dance that indicates distance to the source of forage.

Indication of direction in the dance is made with respect to the position of the sun. In honeybees that nest in the open, e.g., *Apis dorsata* and *A. florea* (Chapter VI – 25), the dances are made directly in respect of the position of the sun. The dancer indicates the direction to the forage by angling the straight run part of the dance at some angle to the position of the sun. Thus, if the returning forager is indicating that the forage is directly away from the present position of the sun, she dances the straight run with her head away from the sun. To indicate that the forage is 90° to the right of the present position of the sun, she dances so that the straight run is oriented at 90° to the right of the present position of the sun. For cavity-nesting species (e.g., *Apis mellifera* and *Apis cerana*), once they are inside the hive, the sun is invisible to them. These bees have made a remarkable leap in communication and have substituted the direction of the downward force of gravity to indicate direction away from the present position of the sun outside the hive! The direction to locations of forage is indicated by **angular deviations** of the straight run part of the waggle dances that are clockwise or counter-clockwise from vertically upward. The simplest example is the straight run of the dancer indicating forage is in the direction of the present position of the sun: she makes her straight run vertically upward on the comb in the pitch blackness of the inside of the hive. Figure 17-4 illustrates several directions to distant locations for forage.

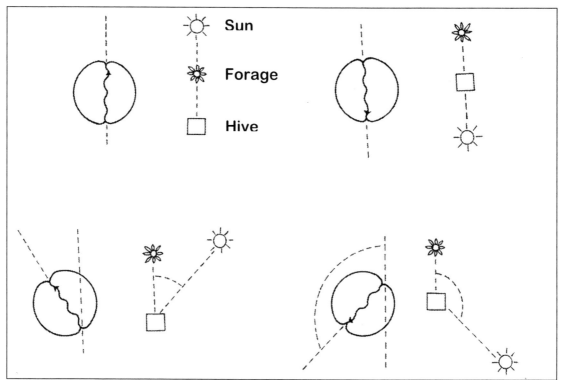

Figure 17-4. Relationship between the direction to a source of forage (depicted by the flower) from the nest, relative to the direction of the sun and the direction of dancing during the straight run of the waggle dance. **Top left:** when the source of forage is in the same direction as the sun, during the straight run portion of the dance, the bees "face" vertically upward on the comb. **Top right:** the bees "face" vertically downward when the source is in the opposite direction. **Bottom left and right:** when the source is 40° or 130° to the left of the sun's direction, the bee dances the straight run at an angle counter-clockwise from the top of the comb.

Other information about the source of forage is also communicated during the dance. The quality and overall **profitability of sources** are expressed by the number, vigour, and duration of dances. One can understand that the stately dance tempo of two circuits per 15 seconds is less exciting than a more vigorous dance indicating much closer forage. Foraging out over 7 km, and back the same distance, is not likely to be energy efficient (Chapter V – 18). The **smells of flowers** adsorb onto the waxy layer of the cuticle and so are returned to the hive and can be a cue that is dispersed during dancing. The taste of nectar and pollen are probably also dispersed as chemical cues. The **time of the day** that resources are available is information that is also returned to the hive and remembered by foragers (Chapter VII – 30).

Just as there are variations in other aspects of Dance Communication (e.g., the distance at which the round dance gives way to transitional dances, the nature of the transitional dances, and the distance at which they give way to the waggle dance) there are **dialects** in the waggle dance. Figure 17-5 presents some of the variations found between species and races in the most distance-sensitive range of the **dance tempo**.

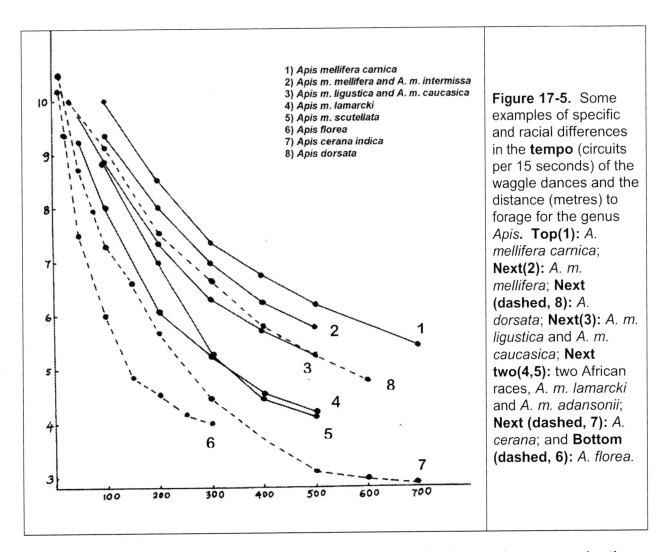

1) *Apis mellifera carnica*
2) *Apis m. mellifera and A. m. intermissa*
3) *Apis m. ligustica and A. m. caucasica*
4) *Apis m. lamarcki*
5) *Apis m. scutellata*
6) *Apis florea*
7) *Apis cerana indica*
8) *Apis dorsata*

Figure 17-5. Some examples of specific and racial differences in the **tempo** (circuits per 15 seconds) of the waggle dances and the distance (metres) to forage for the genus *Apis*. **Top(1):** *A. mellifera carnica*; **Next(2):** *A. m. mellifera*; **Next (dashed, 8):** *A. dorsata*; **Next(3):** *A. m. ligustica* and *A. m. caucasica*; **Next two(4,5):** two African races, *A. m. lamarcki* and *A. m. adansonii*; **Next (dashed, 7):** *A. cerana*; and **Bottom (dashed, 6):** *A. florea*.

The dance language is accompanied by an **acoustical** or **sonic communication**. The sense of hearing in honeybees is not well understood, and for many years it was thought that they had none. In fact they have the ability to detect sounds below the range heard by human ears.

During dancing, a returned forager beats her wings, making sounds that travel through the air and in concert with her body (dance) language. Those sounds (ca. 250–320 Hz), faint to the human ear, are audible to other bees through their Johnston's organs on the antennae (Chapters I – 2 & VII – 28). The dance follower also produces sound, but she does so with thoracic vibrations that she transmits through the comb by pressing her thorax to it. The message through the comb causes the dancer to stop and give out a sample of what she has brought back to the hive.

119

Chapter 17
Dance Communication

Sonic, or vibrational, communication is important in stingless bees (Meliponini). These bees make sounds near the entrance of their nests. Those sounds are interpreted by nest-mates who use the information as part of their orientation to a source of forage. Different species have different signals that combine dance and sound in various ways. James Nieh is unravelling the mysteries of communication in stingless bees. Nieh and co-workers have suggested that complex communication within the hive may have arisen to avoid espionage, i.e., different species and different nests stealing information from each other about the location of resources!

In foraging communication, honeybees and stingless bees use the body and acoustical components of the dance language. They also use a wide variety of other cues. They are able to determine the position of the sun even on cloudy days through their appreciation of polarized daylight (Chapter VII – 26). They remember visual and olfactory features of the landscape (Chapter VII – 27). They are able to follow each other through watching and tracking by scent. Once close to the source of forage, they use colour vision, pattern and size recognition, movement, surface texture, and olfactory cues, including pheromones, to land, find the forage, and gather it (Section VII).

Even though the dance language is generally accepted as an integral part of honeybee (and stingless bee) biology, it has been brought into question. Even the most appealing of explanations can be questioned if examined in detail. The initiators of the **dance language controversy**, Drs. A. M. Wenner and P. H. Wells (1990) pointed out in the 1960s that honeybees returned with the scent of where they had been on their bodies, often left pheromones at the food source, could see and follow each other, and can remember visible and scent characteristics in the general landscape. They also noted that the distance information given in the dance was not fully accurate, and that no similar system of communication seemed to be represented in other social Hymenoptera, e.g., hornets, yellow jackets, and bumblebees.

Although Wenner and Wells conducted a series of experiments which indicated that honeybees do not always follow the instructions conveyed by the dance, they did not disprove that the dance was interpretable.

In one experiment, they arranged two hives of bees side by side. They kept one hive closed so that the bees could not forage, and they set up a feeding station to which the open hive had access. Of course, all the bees foraging from the open hive went to that station for forage (sugar syrup), although we can assume that some "scout bees" (Chapter V – 18) were abroad, looking elsewhere. The researchers then placed other feeding stations in the experimental area and opened the other hive. The foraging bees from both hives visited all the feeding stations in much the same proportions. The originally open station, farthest of all from the hives, was visited the least. In another experiment, with two hives well separated from each other, the same tactic was used: one feeding station, one hive open, the other hive closed followed by several feeding stations (in line from the open hive to the original feeding station) being opened and the previously closed hive now being allowed to forage. The result again indicated that the bees distributed themselves amongst all the stations, but the proportions of foragers from each hive seemed to distribute themselves among the stations roughly according to distance. The important result was that the foragers from the hive that was open throughout the experiment changed their attention dramatically from the originally open single feeding station. That suggested to the researchers that the dance language was not being used.

The results of other experiments have shown the validity of von Frisch's original interpretations of the dance language. In an experiment involving four feeding stations and over 4,000 individually marked honeybees, students at the California Institute of Technology showed that 91% of the bees following a particular returning forager's dance arrived, within 4 minutes, at the feeding station she had indicated, which was ca. 150 m from the hive. When the quality of the food at the station was varied between rich syrup and weak syrup, 93% of the dances indicated the station with the rich resource, regardless of whether or not scent was added to the syrup.

In a particularly elaborate experiment, "fibber" bees were created. Into the experimental hive, an artificial sun (small, dim light bulb) was placed on a curved rail so that its position "in the sky" could be varied. Some bees in the hive were "ocelli-blinded" by aluminium paint. In the hive, normal bees could see the artificial sun in the hive, but the ocelli-blind bees could not. Chapter VII – 26 explains that the ocelli are sensitive to dim light, but that the normal compound eyes require bright light to respond. By changing the position of artificial sun inside the hive, bees that could see the artificial sun as well as the real sun outside the hive (i.e., normal bees), and that were recruited by a returned forager, would exit the hive and follow the direction (angular deviation) indicated by the dance with respect to the "false sun". The ocelli-blind recruits, unable to see the "false sun" in the hive, would exit the hive and follow the foragers' directions with respect to the direction of the real sun outside. The recruited "new" foragers were collected at the feeding station (A) from which the "old" forager had returned, or from a station (B) opened in the place indicated by the position of the "false" sun, to collect only "new" recruited foragers. The ocelli-blind bees were collected at station A, but normal bees were collected at station B.

Although the foraging dance language is now well understood, the means by which honeybees **navigate**, **orientate**, and **measure distance** are parts of the fulfillment of foraging. Honeybees use a complex combination of visual, olfactory, and possibly magnetic senses in navigation and orientation (Section VII). They use the **sun compass** for orientation, and compensate for the traverse of the sun in the sky over hours of foraging. On cloudy days, bees sense the position of the sun by their ability to see the angle of polarization of daylight. Moreover, it is well known that honeybees remember and use **visual and olfactory landmarks** to assist in navigation and orientation. The honeybee's senses are well integrated with the effective communication systems of the colony.

How honeybees are able to **measure distance** has been much debated. A variety of experiments have presented clues that suggest that somehow honeybees measure the amount of effort they expend on a foraging trip. Table 17-2 presents some general results of experiments designed to examine how honeybees measure distance.

Table 17-2. Ten Experiments and Their Effect on Distances
Communicated by the Dance Language

	Experiment	Result on Dance Language
1	If weight is added to a bee	distance is overestimated
2	If wind resistance is added to a bee (a small "sail" glued to the top of the thorax)	distance is overestimated
3	If bees have to fly uphill to forage	distance is overestimated
4	If bees have to fly into the wind to forage	distance is overestimated
5	If bees have to fly downhill to forage	distance is underestimated
6	If bees fly with the wind to forage	distance is underestimated
7	If, while a forager is at a feeding station, the station is moved nearer to the hive	distance is overestimated
8	If, while a forager is at a feeding station, the station is moved farther from the hive	distance is underestimated
9	If, when bees are trained to forage at a ground-level feeding station, the station is then gradually elevated on a tower or onto a high bridge over a gorge	the trained bees continue to forage at the station but recruits fly beyond the base of the tower or under the bridge and farther to search for the forage
10	If bees are trained to forage at a feeding station that is beyond the hive and behind a large barrier (e.g., large, tall building)	the dance communication reflects the direction accurately, and reflects the distance as required to fly around the barrier

The conclusion from those 10 experiments in Table 17-2 is that honeybees measure the amount of **effort** expended (in all experiments) on the **outbound** flight (in experiments 3–10) to estimate the distance.

Recently, though the results of downhill and uphill foraging (in experiments 3 and 5) have been questioned. Harald Esch and John Burns conducted an elaborate series of experiments to ask if **"effort"** was the operative measure to explain the determination of distance.

They placed an experimental beehive 158 m from the near top corner of a 50 m high building, and 158 m from the far front bottom corner of the same building. They trained bees to forage at feeding stations on both the top of the building and at the front of the building. If the information given in the dance language by returning foragers was dictated by effort (1.52 versus 1.03 joules expended in flight to the stations), then those foraging at the top of the building would dance more slowly. Both groups of foragers indicated similar distances to the stations (0.57 and 0.58 dance cycles per second; or about 8.5 cycles in 15 seconds). Measurement of effort did not seem operative.

They also set up hives and feeding stations at the base of, and on the roofs of, two buildings separated by 228 m. They discovered that those bees flying from roof-top to roof-top indicated in their dance language a distance of 125 m, but those foraging at ground level indicated 200 m.

Those results, coupled with others from other related experiments, indicate that **optical cues** are used to measure distance. When bees are flying at a high altitude, the angular change in the image of objects passing beneath them is slow compared to that for bees flying close to the ground. Human beings experience the same effect: a tree at the roadside "rushes past" someone riding in a car, but a tree across a field appears to move across one's field of vision much more slowly. The **optical flow theory** explains the results of all 10 experiments, and of the series of 5 experiments presented by Esch and Burns.

What Esch and Burns discovered was that the distance information given by the dance language varies with the height the bee flies above the ground. Thus, on windy days, bees may fly closer to the ground to experience less turbulence and to take advantage of slower wind speeds closer to the ground and stronger winds higher up (Chapter V – 18) on foraging trips. In general, honeybees fly within a few metres of the ground, but exactly how they adjust their altitude remains to be discovered. It is known that insects, and especially honeybees, have excellent stereoscopic vision and can estimate their distance from objects with great accuracy (Chapter VII – 26). Thus, the combination of **optical flow** with **stereopsis** would allow bees to maintain their altitude in flight.

The results of experiment 9 indicates that honeybees do not have a way of communicating up, even though they can learn to fly upward to obtain forage. An experiment in which bees were trained to forage across a 1.2 km wide canyon showed that at first the foragers maintained low altitude above the ground, flying down the canyon on one side, up on the other to the feeding station, and returning by that route. As the foragers became familiar with the landscape and the task, they flew directly across the canyon and back again. The dance language adjusted to the shorter, more direct, route.

Honeybees have several other dances, but they are not involved directly in foraging. The best known of these is the **dorso-ventral abdominal vibrating dance or DVAV dance**. This dance, in which the workers vibrate their abdomens up and down, has several functions, and stimulates various activities: it can regulate **foraging**, it is involved in **synchronization of swarming** and **stimulating nuptial flights** by young queens. Thus, for workers, it stimulates potential foragers to move to the area where the foraging dances are performed. During the time of queen rearing (Chapter V – 20), the adult queen receives much attention by workers performing DVAV while touching her. She may be vibrated by DVAV up to 300 times per hour. However, that high level of stimulation stops a few hours before swarming occurs (Chapter V – 20). Cessation of DVAV by the workers' touching a young queen is part of the stimulation for her to fly out of the nest on a mating flight (Chapter VI – 21).

The **buzzing run** (sometimes also called the breaking dance) is performed by the workers just as a swarm is about to issue from the hive. The workers run, often in groups, while vibrating their wings. As the intensity of the buzzing runs increase, workers are stimulated to join in and to be led from the hive to fly as a swarm. The same, or similar, behaviour may be involved in stimulating a swarm to settle.

The **tremble dance** has been long known, but its function remained obscure until about 1993. It is a dance performed by returning foragers when there is so much forage being returned to the colony that there is a shortage of workers involved in storage activities (Chapter IV – 15). The dance is accompanied by short, vibrational pulses at about 450 Hz that last about 15 ms (milliseconds). The effects are two-fold. Workers are stimulated to take the forage returned to the hive and store it, and foraging dances and foraging are curtailed. The so-called **spasmodic dance**, also accompanied by abdominal wagging, involves the distribution of resources in the hive.

When a returned forager is about to start dancing, the workers in the vicinity are jostled and pushed aside. **Jostling** seems to alert workers to the incoming news. Workers may **shake** themselves vigorously from side-to-side while in contact with nest-mates. That stimulates nest-mates to groom the shaker, sometimes of sticky propolis (Chapter X – 42), of other foreign matter, or of external parasites (Chapter IX – 35).

In summary, honeybees have a remarkable system of communication. Although they use the news as combined by body, acoustic, olfactory, and gustatory information provided by returning and dancing foragers, they are quite able to exercise choice and adapt their activities according to new information and circumstances. They learn from their own experiences, and adapt their activities accordingly. Honeybees are far from being tiny automatons!

CHAPTER V – 18

FORAGING EFFICIENTLY FOR FOOD

Bees forage mostly for nutritional rewards presented by flowers' **nectar** and **pollen**, sometimes for **honeydew** (Chapter IV – 13), but also for **water** and **propolis**. Some bees forage for oils and gums also presented by flowers. In social bees, such as honeybees (Chapters VI – 24 & 25), stingless bees (Chapter XI – 50), and bumblebees (Chapter VI – 25), foraging is organized hierarchically by integrating behaviours of individual worker bees with colony requirements (Chapter IV – 15). Foraging is part of the social structure, and its ultimate benefit is for the colony rather than for the individual.

Weather and other environmental factors play an important role in allowing bees to forage. For European races of the Western honeybee, foraging takes place only during daylight. It begins when ambient air temperature reaches 12–14°C and light intensity (global solar radiation) is greater than or equal to 0.66 Langleys (a measure of light). Relative humidity is also important as it affects the amount of foraging either positively or negatively, depending on the type of flower foraged: the nectars of some flowers may evaporate to crystals in hot dry weather. Strong winds and heavy rainfall inhibit foraging.

Within colonies of eusocial bees (Chapter I – 1), such as honeybees, **communication** is an integral part of the colony's life as a "super-organism". Foragers gain information about the location and nature of the resource from scouts. These are foragers that explore their environment, looking for resources to be harvested, and return to the colony with information. The way in which scout bees arise, and how they function, is not well understood. It is likely that they fly out from the colony, find a resource, and return with a partially laden honey stomach. This way they can return quickly with the maximum amount of information to convey to their nest-mates. Some foragers may also work as scouts while on dance-directed foraging trips, doing double duty. Other foragers are strictly faithful, constant, to one source of forage, at least on a single trip or set of foraging trips.

The information gathered by honeybee scouts or dance-directed foragers is shared within the colony as a whole rather than with individual bees. That is an important distinction to make because the colony acts as a "super-organism" rather than a collection of individuals. The same idea applies to other animals with highly social behaviour, such as bumblebees, hornets, yellow jackets, ants, termites and so on, even though dance communication is not part of their behaviours.

Animals that forage from a home are called **central place foragers**. Denning mammals, nesting birds, nesting insects (including solitary bees (Chapter VI – 25), and, of course, hive-inhabiting bees are all central place foragers. For all central place foragers, knowledge of home territory is essential, not just for foraging but also for averting risk of catastrophe or death.

Understanding of **central place foraging** and **optimal foraging** (i.e., maximizing the intake or gathering of resources while minimizing the expenditure of energy and the risk of catastrophe or death) in relation to bees has come about through many studies on bumblebees (the genus *Bombus*). They are quite large insects, are easy to watch and follow, and can become used to the presence of curious scientists. Because the colony is somewhat buffered against adversity, individuals may be thought of as somewhat expendable so that risk of an individual's demise is somewhat less important in the larger, colonial, picture.

The total areas used by various bees as foraging territory are highly variable. In some tropical areas, some bees forage over wide ranges, having "trap-lines" or feeding points at particular species of widely dispersed plants over tens of kilometres. Honeybees, especially the larger species such as the Western honeybee (*Apis mellifera*) and the giant or rock honeybee (*A. dorsata*), have large foraging territories. The **median foraging radius** of a colony of European honeybees differs depending on the environment. In agricultural areas, the foraging range may be a few hundred metres to about 3.7 km, but in forested and desert areas the range may be much greater and exceed 6 km. In studying the dance language of colonies of native bees in Sri Lanka, it was found that distances that were preferred, and the maximum communicated, corresponded to the size of the bees. *Apis florea* (the smallest honeybee) preferred to forage within a few hundred metres, *A. cerana* within about 0.5 km, and *A. dorsata* out to about 1 km even though its dance is similar to that of the European races of the Western honeybee (*A. mellifera*).

The constraints on foraging distance and dance communication have implications for beekeeping. Clearly, keeping large numbers of hives of bees in one place would result in the total population of bees having to forage farther afield to satisfy the total nutritional demands of all the colonies. If the dance language can convey information only to short distances (as in *A. florea* and *A. cerana*, Chapter V –17), the colonies in a densely populated apiary would be unable to find enough forage for colonial health. They would starve, or have to abscond (the whole colony would leave as a non-reproductive swarm) (Chapter V – 20 & Section VIII).

Once individual bees find an area in which to forage, the lessons from watching bumblebees point to how efficiency in foraging is achieved. While foraging on the microgeographical scale, i.e., **between patches of flowers in a given area** (Figure 18-1), the worker bumblebee tended to move in one direction, first south and returning north, foraging in both directions. That suggests efficiency. The map shows that the bee tended to follow the same general path, stopping at the same patches of flowers. That pattern (trap-lining) suggests familiarity with environment (learning) and concomitant efficiency. Nevertheless, the bee did vary its path from day to day. That suggests exploratory behaviour (scouting) in looking for other sources of reward while foraging. Finally, the map shows that the bee visited inflorescences of *Aster* but not of *Impatiens*, indicating that the bee specialized on one flower type (**floral constancy**). The elements of efficiency are well demonstrated by this bee in its foraging over an area containing a variety of potentially valuable resources. In other situations, researchers have observed bees testing other potentially valuable sources of reward, lessening their floral constancy in a behaviour that has been termed "majoring and minoring".

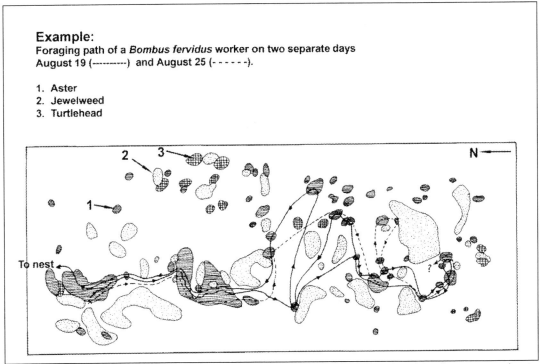

Example:
Foraging path of a *Bombus fervidus* worker on two separate days
August 19 (----------) and August 25 (- - - - - -).

1. Aster
2. Jewelweed
3. Turtlehead

Figure 18-1. Bumblebee foraging paths in an open area of mixed herbaceous plants.

Once on a patch of flowers, a bee should forage in a way that maximizes the amount of resources it gathers while minimizing the energy it expends (i.e., optimal foraging). Figure 18-2 shows a representative foraging path of a bumblebee on a patch of flowers.

What patterns of foraging are shown? The bee flew more or less in one direction. This forwarding pattern suggests efficiency, especially because she would be unlikely to cross her own path and encounter a flower she had already visited. The bee flew more or less into the wind. That behaviour suggests she used the wind to aerodynamic advantage in takeoffs and controlled landings. The bee mostly alternated left and right turns unless she flew straight ahead. That pattern enforces the forward motion of the bee within the patch, yet allows for visiting many flowers over a wide band.

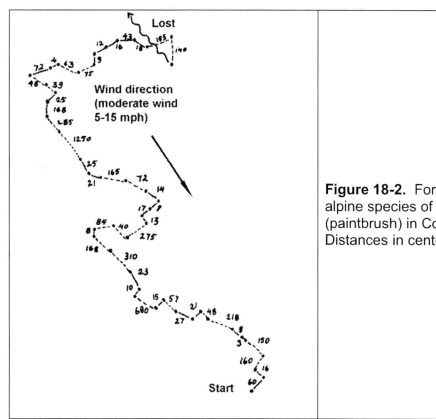

Figure 18-2. Foraging map for a worker of an alpine species of bumblebee on *Castilleja* (paintbrush) in Colorado, July 31, 1981. Distances in centimetres, not drawn to scale.

A bee, foraging on a patch of flowers, must be able to assess the worth of the patch. How does a foraging bee forage on a patch of flowers that offers a rich supply of reward versus a patch that is poor?

It would be efficient for the forager to remain in a rich patch, and abandon a poor one. Bees have built-in behaviour patterns that enforce such behaviour. If the patch is poor, foragers move mostly straight ahead and fly longer distances between flower visits. On the other hand, if the patch is rich, foragers make many more sharp turns to the left or right and fly shorter distances between flower visits. The results of some experiments show that bees conform to those "rules" of foraging.

The effect of the amount of nectar present in flowers and the distances foragers fly to the next inflorescence is clearly shown by the bar chart in Figure 18-3 (right). In another experiment, a patch of flowers with high nectar reward was created by placing netting over it to prevent bees from foraging there, and allowing the nectar to accumulate to a level of about 0.01 mg of sugar per flower. The relatively poorly rewarding, and adjacent, patch was open to foraging, and the flowers had only 0.003 mg of sugar per flower. The left-hand graphs in Figure 18-4 (A) and (B) show that the bees tended to "forward" and skip over many flower heads when nectar rewards were low. The right-hand graphs in (A) and (B) show that bees did not move in the same direction (forward) after successively visiting flowers that contained rich rewards.

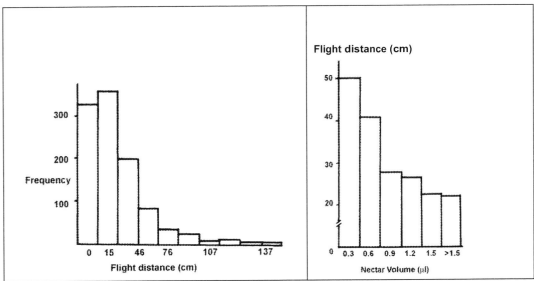

Figure 18-3. Left: The general pattern of flight distances of bumblebees (*Bombus pennsylvanicus*) foraging between successively visited inflorescences of *Delphinium virescens*. **Right:** The mean distances flown by the same species of bumblebees after visiting inflorescences with the indicated mean volume of nectar artificially injected into the two lowest flowers. In nature and to foraging bumblebees, the two lowest flowers have been found to be good predictors of rewards in other flowers on the same inflorescence.

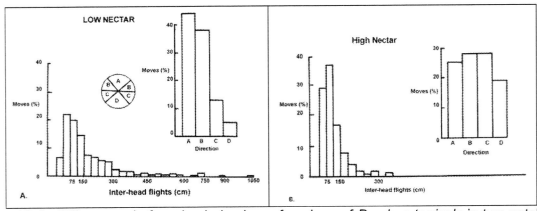

Figure 18-4. Differences in foraging behaviour of workers of *Bombus terricola* in two patches of white clover: (**A**) Low nectar reward (a poor patch), and (**B**) High nectar reward (a rich patch). The distances flown between flowering heads of white clover are shown by the bar charts on the left. The possible directions of movement are as shown in the circular figure in which A is forward movement and D is backward, and the frequency with which the bees chose to move in any of those directions is shown on the right-hand bar graphs.

Figure 18-5 illustrates the net effect of foraging according to the two rules for distance and direction. In both hypothetical patches, the point sources of reward are represented by dots and are identically placed. The forager enters and exits the patch in exactly the same place, but by visiting point sources that are close together and turning sharply to left or right, the forager in the rich patch visits four times as many point sources.

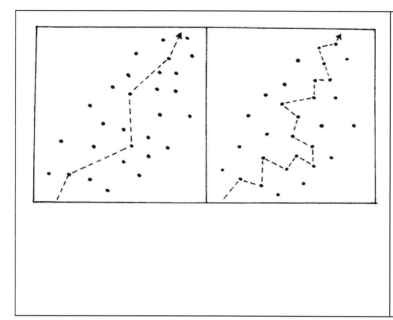

Figure 18-5. The net effect of stopping at closely spaced point sources of resources (flowers with nectar or pollen) and turning sharply left or right between successive visits keeps the forager in the rich patch longer so it visits more flowers. Thus, it forages efficiently where there is an abundance of food. On the other hand, passing quickly and more or less straight through a patch of limited food, the forager samples and checks as it progresses, thereby monitoring the levels of resource in various patches within its foraging territory.

Once a foraging bee is foraging at an **array of flowers on a plant**, e.g., flowers in an inflorescence, on a bush, or on a tree, more efficiencies come into play. Generally, bees foraging on arrays of flowers, such as of fox-glove (*Digitalis* spp.) (Figure 18-6), fireweed (*Chamaerion*), *Delphinium*, monk's hood (*Aconitum*), etc., start low and work upward. In nature and to foraging bumblebees, the two lowest flowers of an inflorescence are good predictors of rewards in other flowers on the same inflorescence.

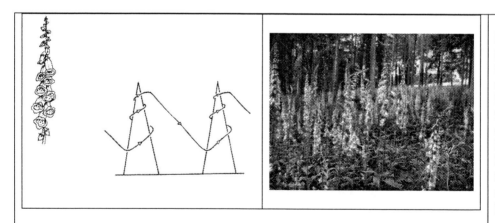

Figure 18-6. The flight path of bees visiting flowering spikes (one of several types of inflorescence). The forager starts low, ascends, then, between inflorescences, drops to the lower flowers and repeats the pattern.

There are several reasons why that pattern of movement is efficient. From an energetic point of view, short scrambling, aided with some flight while ascending, requires less energy than would a steep ascending flight between inflorescences. There is also the matter of the ease and accuracy of landing: slowing to land with the help of gravity, and pitched head-high allows rapid stops with the correct body orientation on a flower. It has also been documented that nectar in the lower flowers is a little more watery, but more plentiful, than in the higher flowers. Thus, as the forager ascends, the taste of the nectar becomes sweeter, off-setting gustatory satiation. One can liken it to people at a wine or cheese tasting: one starts with the more delicately flavoured wines or milder cheeses which would be difficult to appreciate after tasting full-bodied wines and strong cheeses. From the botanical side, generally the flowers at the tip of an inflorescence are the youngest and are shedding pollen; the stigmas are not yet receptive. Older flowers are in their female phase, with the pollen mostly gone and the stigmas receptive. Thus, the forager moves pollen from the younger, top flowers, to the older, sexually receptive lower flowers of another inflorescence. The bee's behaviour, coupled with the maturation of the flowers in the inflorescence, promotes cross-pollination (Chapter X – 45).

Most of the research on foraging movements by bees has been done in meadows and on flowers in vertically arranged inflorescences. Some research done on patterns of bees' movements on small, bushy plants suggest no particular systematic foraging behaviour. However, on trees, large bees follow foraging patterns (Figure 18-7).

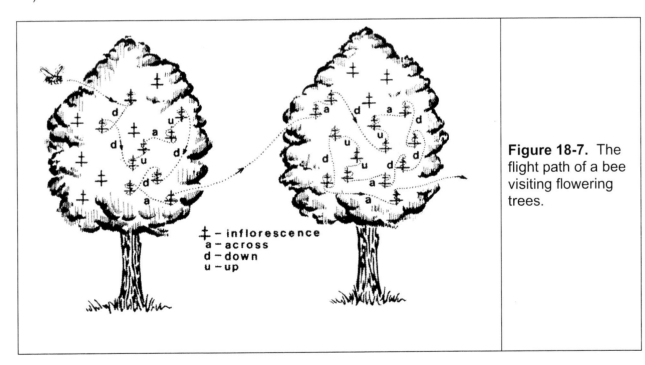

Figure 18-7. The flight path of a bee visiting flowering trees.

The pattern of foraging on inflorescences on trees (e.g., horse chestnut (*Aesculus* spp.), black locust (*Robinia*)) is the same as already described, but the foragers treat inflorescences that are vertically close together (e.g., within a few centimetres) as a single inflorescence and so continue their upward course over several inflorescences. They treat separated inflorescences in the same way as inflorescences on the ground, but tend to descend farther than the level of the base of the inflorescence they have just left. Thus, depending on the density of inflorescences, bees may slowly ascend trees, slowly descend, or remain at more or less the same level as they forage. Mostly, it seems that the density of inflorescences on the surface of the crown of a tree is such that bees leave the tree at a lower level than they enter it. This pattern of foraging is also efficient: the bees enter high and exit lower (with a greater load of harvested resource). Why is this?

One can imagine being given the task of removing bags of gold coins (nectar or pollen) from floors of a tower (inflorescences on a tree). If the average weight of treasure in the tower is 5 kg per floor, and there are 10 floors to the tower, it is simple to understand that by climbing to the top of the tower and gathering the bags on descent, one is carrying weight down (rather than up) and one emerges at the base of the tower with 50 kg of treasure gathered in the most efficient way.

Learning is an integral part of a bee's life. Bees can learn to recognize physical and chemical components of the landscape over which they forage. They learn foraging routes, and learn to associate various features with the presence of resources, such as nectar, sugar syrup, pollen, pollen substitute, and so on. Although bees have innate, non-learned, instantaneous responses to certain stimuli, learning is not instantaneous. Learning requires trial and error, and an investment of time while reactions to particular stimuli or circumstances are honed. Figure 18-8 presents the standard sort of learning curve experienced by all animals that can learn. In this curve, honeybees have been trained to associate obtaining a reward with a task that requires learning. As the number of experiences increases, so does the accuracy of choice.

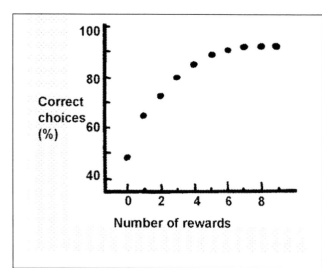

Figure 18-8. A honeybee worker makes increasingly accurate choices the more times she is exposed to the experience and obtains a reward.

Flowers come in a stunning array of shapes and sizes. Open bowl-shaped flowers mostly expose the forage that bees seek. More complex, tubular flowers have hidden nectar, and sometimes pollen. In even more complex flowers, nectar and pollen are concealed so that a forager must cope with the floral complexity if she is to obtain forage. The botanical side to all that is the placement of the flower's sexual parts, the anthers and stigmas, which must touch the foraging pollinator for transfer of pollen to stigma. In simple flowers, the sexual parts are exposed and accessible, but in complex flowers, they are hidden and rather intricate in their placement and operation. The placement of the floral rewards, such as nectar, pollen, and oil, requires the foraging pollinator to work flowers so as to bring about pollination.

Terence Laverty conducted a number of elegant experiments in which he showed the capacities of bumblebees to learn to manipulate flowers. He showed that worker bumblebees can manipulate simple flowers (open bowls or simple tubular flowers) and obtain reward without practice. Bees have an innate reaction to flowers, and attempt to forage from them even though they may never have encountered a flower before. It is simple to obtain naïve bumblebees by keeping colonies (Chapter XI – 48) in artificial surroundings. The colonies can be exposed to real or artificial flowers for many kinds of experiments. Laverty found that naïve bumblebees required practice before they could forage from slightly complex flowers, such as *Prunella* and *Apocynum* (dog bane). When confronted with highly complex flowers, such as those of turtle head (*Chelone*), the bees required much more practice to be accurate and efficient (Figure 18-9).

Laverty also showed that some species of bumblebees have greater capacities to learn to manipulate complex flowers with which they are known to be associated as specialists. Specialization of foragers of honeybees, bumblebees, and stingless bees for flowers is well known, but the way in which individuals assume specialist roles is not.

In general, larger, complex flowers secrete much more nectar than do simple, open flowers. The differences may be 100-fold or more. Thus, investment in learning (education) would appear to have its benefits through greater access to more resources. (Students, take note!) The question arises as to why all bees from a colony do not become specialists. It must be presumed that from the perspective of the colony as a super-organism, it is more efficient to invest in as many learned specialists as are needed to take the greatest advantage of large amounts of reward at complex point sources (flowers), which may be relatively few, but also to have a labour force that requires less training and that accesses point sources with easily accessible but small amounts of reward. Moreover, it may be that some workers are more adept at learning than others. Scientists who train bees to learn about their capacity for sensory discrimination through associative learning have long known that some individuals never attain proficiency: perhaps they are stupid.

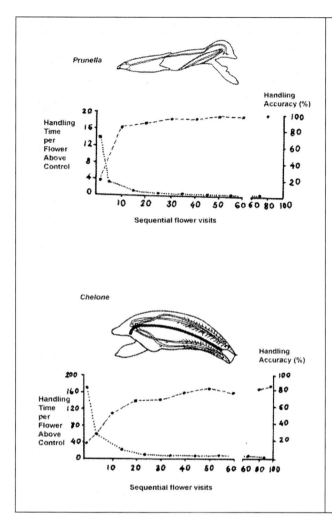

Figure 18-9. The diagrams demonstrate increased accuracy and decreased handling time in foraging efficiency at simple and complex flowers in sequential visits.

This chapter started at the level of foraging over the landscape, finding and using patches of resources. Honeybees and stingless bees have refined systems of communication (Chapter V – 17) to facilitate efficiency in foraging, an activity that is largely determined by a worker's age (Chapter IV – 15). The colony acts as an "information-centre". The colony as a whole, as mediated through its individual members, has the capacity to compare information about all of the patches discovered by scout worker bees and returning foragers. Presumably aspects of the dance language (e.g., vigour and numbers of identical dances taking place at any given time) determine which patches are the most profitable. Once a richer source of food is discovered, the colony's activities switch. The initial "news" may be efficiently brought to the nest by scouts with only partially laden crops, perhaps as part of a suite of behavioural adaptations that increase the time spent in the colony and in turn increase the information centre functions of the nest. Beekeepers know that in any given apiary, the conspicuous directions taken by foragers leaving from, or arriving at, any particular hive may not be the same as at another hive. That indicates that different colonies express different "opinions" as to the best places to forage.

Although bumblebees appear to lack anything akin to dance communication, as individuals working in the interests of their colony, they track the location and quality of the resources they need to remarkable effect. All female bees are central place foragers and so must be able to orient themselves to their nests. They must also find and learn the location of the forage they need to provide for their offspring (Chapter VI – 25).

This chapter also started with the topic of **optimal foraging** (i.e., maximizing the intake or gathering of resources while minimizing the expenditure of energy and the risk of catastrophe or death), so it is worthwhile to close with a consideration of whether or not bees optimally forage. There are four assumptions about optimal foraging that apply in evolutionary ecology as well as in practical beekeeping:

1) An individual's contribution to the next generation, known as "**Darwinian fitness**", should be enhanced by more efficient foraging and minimized risk through resistance or tolerance to disease, parasites, pests, and predators. That is true when one takes the view that the colony is an individual super-organism. The more resources that are stored, the healthier and more prolific is the colony. Such colonies are favoured by beekeepers and used in selection and breeding, although predator resistance as indicated by defensive behaviour is not so valued.

2) A component of foraging behaviour is heritable. It is known that honeybee colonies can be selected for high hoarding behaviour, and for foraging at flowers of certain plants, e.g., alfalfa (Chapter VI – 23). In other bees, notably bumblebees, there are interspecific differences in foraging behaviours.

3) There are "functional" constraints to the extent to which optimal foraging can be expressed. Different species of honeybee, stingless bee, bumblebee, and so on differ in the form of their mouthparts (length), legs, eyes, and so on, and those features limit foraging ability: some efficiencies may be gained, but others lost (Chapters VI – 23 to 25). Even the ability to learn differs subtly between species of bumblebees.

4) The final feature of optimal foraging, often mentioned, is that the evolution of traits promoting it should be rapid. It is not clear why that should be, but one would expect evolution for optimal foraging to be directional and driven by selection, natural in most cases, and through breeding and selection in beekeeping.

CHAPTER V – 19

STAYING WARM & BEING COOL: THERMOREGULATION IN HONEYBEE COLONIES

The maintenance of body temperature in animals is a result of the **process of metabolism** (the chemical changes and processes of living cells). The temperature of cold-blooded animals, such as insects, amphibia, reptiles, and most fish varies with the temperature of their surroundings. In general, their temperature is always slightly below the outside temperature to prevent the loss of body moisture through evaporation. For warm-blooded animals, metabolism must be maintained at a high rate to generate the heat required for the functioning of a complex brain and nervous system, and to maintain muscle tone. In warm-blooded animals, the cells in the body operate most efficiently within a narrow range of temperatures. If body temperature is too high or too low, the cells themselves become damaged. The processes of thermoregulation, such as shivering, sweating, and panting, are controlled involuntarily by the brain.

Many social insects, such as termites, ants, wasps, bees (including bumblebees and honeybees) **thermoregulate** in their nests. Colonies of cavity-nesting honeybees (*Apis* spp.) are probably the only insects that can maintain a temperature above that of the air outside throughout the winter in temperate countries. Similarly, they are able to keep the nest cool during especially hot weather. It is likely that all species of honeybees are adept at thermoregulation within certain ranges of environmental conditions.

In cavity-nesting honeybees, the site and architecture of the nest plays an important role in moderating temperature extremes. The choice of **nesting site** by a swarm of honeybees is made with regard to **exposure** to wind, sun, and rain. When swarms (Chapter V – 20) choose a cavity for occupation, overly wet and drafty ones are rejected. If the cavity being considered by the swarm has some faults, workers can plug holes and seal damp patches with propolis (Chapter X – 42) to make the cavity more habitable. Experiments with artificial cavities (boxes or hives) of different sizes and orientations have demonstrated the preferences of European honeybees. The average **nest volume** they choose is about 40 L (within the range of 20–100 L).

The **orientation and size of the entrance** is important for summer ventilation and winter heat retention. Cavities occupied by European honeybees in northern temperate countries usually have south-facing entrances because they are sunnier, warmer and drier than if they faced another direction. The **size of entrance** is generally smaller than 60 cm^2. On average, the entrance to a natural nest of European honeybees can be found 3 m above ground.

Within a natural cavity, the nest architecture embraces marvels of natural civil engineering (Figure 19-1).

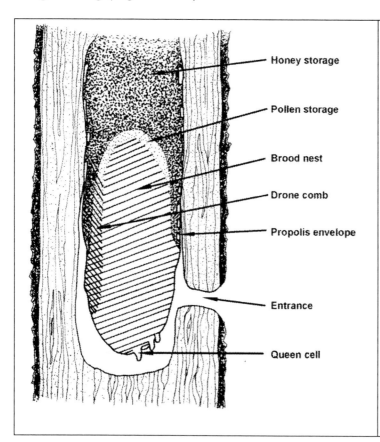

Honey storage

Pollen storage

Brood nest

Drone comb

Propolis envelope

Entrance

Queen cell

Figure 19-1. Longitudinal section of the nest of a natural colony of honeybees inside a hollow tree.

The **propolis lining** reduces heat flow out of the cavity and acts as waterproofing. Propolis also has strong anti-microbial qualities and can protect the colony from some infections (Chapter X – 42). The design of **comb construction**, the hexagonal cells, contributes also to the integrity of the nest. The heaviest stores (the honey) are placed at the top of the nest. Thus, in winter the honey is warmed by convected air rising from the cluster. That reduces its viscosity and tendency to granulate (Chapter X – 38) so that it is as easy as possible to use. The points of attachment of the series of parallel combs in the cavity are close to the mass of honey so that in a cavity in a tree swaying in a storm, the comb is less able to swing back and forth than if the honey were freely suspended to create a pendulum effect. The brood area is in the airiest part of the cavity. There, when needed, it can be provided with fresh air drawn in through the entrance. The pollen stores are usually between the brood and the honey. The entrance is large enough to accommodate the comings and goings of foragers, and provide space for the guard bees, in summer. It is small enough that it can be defended against intruders, and even closed partially in winter by a propolis curtain. The cells for rearing the sexual castes are found in particular areas, too. Queen cells, looking somewhat like large peanuts, hang from the bottom of the combs for the brood nest. Drone brood cells are mostly found in the brood area of the combs, but at the back, away from the entrance of the hive.

Although the cavity and its contents are well constructed to protect the colony, how does thermoregulation operate in the life of a colony? Honeybees, stingless bees, and bumblebees **incubate** their brood with great precision. To illustrate the importance of incubation, the capped brood of those bees can be artificially incubated because the larvae no longer need to be fed (Chapter IV – 12). The temperature at which the incubators are kept is as important to the bees as to birds' eggs or human babies.

For example, when capped brood of European honeybees and *Apis dorsata* are kept in incubators at different temperatures, the results are startling. At 26°C, almost no adults emerge, and those that do are badly deformed and take longer than normal: at temperatures between 28 and 30°C, almost all the brood ecloses (emerges as adult workers) after a short delay, but the adult bees have shrivelled wings and malformed mouthparts. Only between **32 and 36°C do normal bees eclose on schedule**. Then, at temperatures just 2°C more, and higher, all the brood perishes!

The reason for the colony's maintaining tight control on the temperature of the brood nest is clear! The colony is acting like a warm-blooded animal in rearing and incubating its young.

A colony of bees does not always have brood (Chapter V – 20). In north temperate areas, brood rearing declines in the fall, and often ceases in the depths of winter (December and January). Even so, the colony must stay active and warm, but without need for such tight thermoregulation.

Individual honeybees have extraordinary capacity for generating heat, but can be active over a wide range of temperatures. At an ambient air temperature of about 14–15°C, an individual honeybee becomes comatose and immobile if she does not exercise her muscles (**chill-coma** or **cold torpor**). She does not die unless kept so chilled for several days. At 7–8°C, an individual worker can keep herself warm, be active, and walk around, albeit slowly. Once ambient air temperatures reach about 16–18°C, she can activate her flight muscles and forage. The worker honeybee's minimum rate of metabolism occurs when her body temperature is about 10°C, the temperature at which she has the **minimum respiratory rate**.

To keep honeybees over winter in special "overwintering houses", as is done in especially cold places such as the Canadian prairies, beekeepers must understand winter metabolism, clustering, and thermoregulation is important. Although the bees are active in the hives inside the overwintering house, they consume stored honey.

Overwintering houses kept at about 4°C provide an environment that keeps the colonies' metabolic, and food consumption, rates as low as possible.

Metabolic rate is measured as microlitres (μL) of oxygen (O_2) consumed per bee per minute. A cool honeybee at about 10°C has a resting metabolic rate of about 2.87 μL O_2/min/bee. She does not stay with that body temperature for long, but when she **generates heat** and elevates her thoracic temperature to about 38°C, which she can do in less than 4 minutes, there is no visible sign of activity but her metabolic rate soars more than 50-fold to about 150 μL O_2/min/bee. A flying worker uses about 460 μL O_2/min. Although during the process of elevating her temperature the worker appears not to change her activity and may seem at rest, in fact her thoracic muscles are shivering rapidly and so generate heat. Special instrumentation is used to detect the vibrations of the thorax itself.

Thermoregulatory activity differs according to the activities in the colony. Keeping warm in winter in a **broodless cluster** requires less heat and less energy consumption than the maintaining of carefully regulated temperatures **when brood is to be incubated**. Thermoregulation **in summer** sometimes involves individuals keeping cool and preventing the temperature of the brood from exceeding the lethal limit.

At any time of the year, or at night, workers begin to cluster when the ambient temperature in the hive declines to about 18°C. Even so, many workers continue activities outside the cluster. When the temperature drops to about 14°C and lower, the bees form a compact sheath of workers as insulation on the outside of the cluster. The cluster is clearly delineated by this sheath or layer of workers, generally several bees thick, oriented with the heads directed toward the centre, especially when it is very cold. Few workers can be found outside the cluster. Inside the sheath is the inner, looser core of moving bees. Throughout that core, the bees are continually moving, creating passages for air movement, and for their own travel internally. By forming a cluster, the colony greatly reduces (by about 30-fold in a cluster of 15,000 bees versus about 2 cm^2 per separate bee) its total surface area exposed to cooling, retains heat, and maintains internal currents of warmed air. The colony becomes its own insulated central heating unit. The energy used to generate the heat, and fuel the bees' activities in and on the cluster, is derived from consumption of stored honey (Chapter IV – 13).

The cluster has internal regulatory facility to react to changes in ambient air temperatures. As temperatures drop below 14°C, the cluster tightens until at –5°C it has become as small and dense as it can (Figure 19-2). At that point, the workers can generate heat only for adjusting temperature control. At temperatures below about 8–10°C, it is rare to see bees away from the cluster. They quickly cool to the temperature of the ambient air. That impairs their mobility, and they may become immobile (chill-coma or cold torpor). Cold torpor starts to immobilize bees once their body temperatures chill to about 14 - 15°C.

Rearing brood requires **stable conditions**. The temperature in the brood nest is held at about 35°C no matter what the outside temperature is. A thin layer of workers covers the brood area to generate heat. There, the workers adjust their metabolic rates to accommodate the temperature needs of the capped brood. Other parts of the hive are allowed to fluctuate, within limits, with the ambient air temperature (Figure 19-3). Despite the cold in mid to late winter, brood rearing starts even in the coldest places where beekeeping is practiced.

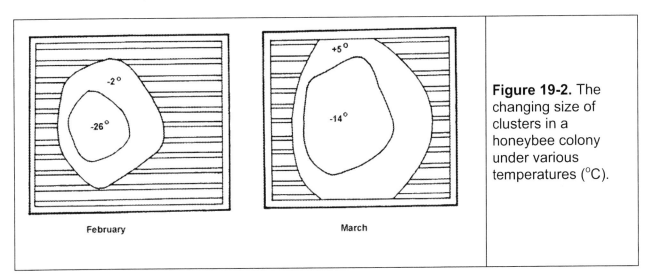

Figure 19-2. The changing size of clusters in a honeybee colony under various temperatures (°C).

The interaction of cluster density, porosity, and thermogenesis results in the colony's having a **thermoneutral zone** from about +10 to –10°C, over which the metabolic rate of the cluster is minimal and stable. At temperatures below the thermal neutral zone, metabolic rate increases linearly, even to temperatures as low as –70°C. As temperatures rise above the thermal neutral zone, metabolic rates in the cluster increase slowly, probably in response to there being some need for cooling and ventilation. A thermoneutral zone is characteristic of hot-bloodedness, or maintaining a constant body temperature (homoeothermy). Human beings have a stable resting metabolic rate over a temperature range of about 20–30°C.

Long periods of cold weather can have a serious effect on hive survival. It can be difficult for the cluster to balance the needs of incubating brood with the need to feed on honey and pollen stores. The winter cluster must include comb for brood and must contain honey and pollen. Protracted, intense cold prevents clustered bees from moving from one area of the hive to another when supplies of honey become exhausted in the cluster's location. Short spells of mild weather, when the temperature in the hive can exceed about 10°C, allow the workers to **break cluster** temporarily to fetch food from another part of the hive, or for the cluster itself to move. If prolonged and intense cold prevents the bees from breaking cluster, they may starve *in situ*, even though stores may be present only a short distance away in their hive. **Cold starvation** is indicated by dead workers, lying head inwards, in the comb. Proper overwintering management is essential in parts of the world with cold winters (Chapter VIII – 32).

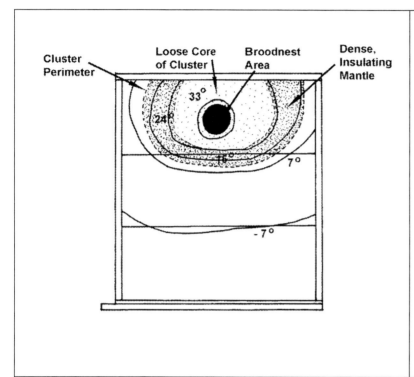

Figure 19-3. Anatomy of a winter cluster of honeybees. In the outer shell, the body temperature of a single bee is close to ambient temperature. Young bees seem not to be able to raise their body temperatures easily. They are kept warm in the brood area, where their age-related activities (Chapter IV – 15) are most needed. Adult bees are the main source of heat for the colony, and are thought to comprise the sheath.

Heat stress on honeybee colonies can be a major problem in hot climates and in hot weather. After all, it takes only a few degrees above the incubation temperature of 36°C to kill the brood. The tolerable duration of extreme heat is not known.

Honeybees can lower the temperature in the hive by spreading out within and ventilating the nest by fanning. They also forage for water that is spread on the combs throughout the hive by younger bees. Water foragers returning are met enthusiastically. The water is spread in tiny puddles on capped honey and brood cells, and suspended as droplets over brood cells. The fanning and ventilation evaporates the water, thereby cooling the colony in the same way as an air-conditioning unit removes heat from a building. Fanners arrange themselves so as to drive air along existing unidirectional air currents, and those at the entrance face inward and expel air. Some bees, e.g. *Apis cerana* are also thought to pump air into their hives. It has been estimated that 12 fanners can move about 1 L of air through a hive in a second (i.e. 60L/min). That is a remarkable feat of natural bioengineering for air conditioning.

Under some circumstances, especially on hot summer nights when all the foragers are also at home, workers may spend the night on the outside of the hive.

If heat stress is extreme, as for example when a hive is threatened by fire, the bees make every effort to cool the nest, even though eventually the wax melts (Chapter X – 39) and the colony is serious trouble.

For keeping cavity-nesting bees in the tropics (*A. mellifera*, *A. cerana*, some Meliponini) (Chapters VI – 24 & 25), hive design and shading are important. Hives with thin walls and roofs are poorly insulated against the heat of the warmest times of day, and against the effects of direct insolation. Hives made of thick, heat-buffering materials, such as coconut trunks, concrete, or clay, or those kept in building walls, seem to hold their resident colonies. Free-standing hives support healthier, stronger, and more productive colonies when they are shaded by vegetation or a roof.

Thermoregulation can be an amazing **defence strategy**. The Asiatic hive bee, *A. cerana*, has evolved an effective defence against several species of wasp (*Vespa* spp.) that can invade their hives and destroy the colonies. Those wasps are huge, by comparison with workers of the honeybees. They are voracious predators of larval, pupal and adult honeybees. It is fascinating to watch foragers evade capture while working flowers. They monitor the position of the predatory wasp, as the wasp watches the bees. Both potential prey and predator fly and hover in an aerial ballet, the bee snatching quick visits to flowers and the wasp darting at the bee.

At the hive, invading wasps can have serious effects. Often, though, the bees dispatch the depredatory invaders by **cooking them to death**. Up to 100 worker bees make a defensive ball over the invading wasp and increase their body temperatures to 48–50°C. That, over a few minutes, is lethal to the wasp.

Another strange phenomenon involving **thermoregulation** and bees is associated with **yellow rain**. Yellow rain was reported to be an agent of biological warfare in Southeast Asia during the time of the Vietnam hostilities. It turned out that the yellow rain was bee faeces (Chapters II – 5 & IV – 13) from the giant or rock honey bee, *Apis dorsata*. These bees make single, large combs suspended from limbs of tall trees, under eaves, bridges, or other buildings. Periodically, they defecate during massed flights. Often, all the bees from a colony, and all the bees from a number of colonies, all inhabiting one structure (e.g., a tree) have mass defecation episodes within a short period. When that happens, being beneath the bees is noticeable!

Why mass defecation should occur was not understood until it was noticed that "yellow rain" occurred when the thermal stress was the greatest. When the air temperature is at 26°C (the daily mean in Malaysia where the studies were conducted), a worker bee's body temperature is 4.6°C higher. When the air temperature is less than or greater than 26°C, the bee's body temperature is more than 4.6°C warmer than the ambient air. The brood-rearing temperature for *A. dorsata*, at 33.5°C, is similar to that for *A. mellifera*. So, when air temperatures rise to between 28.5 and 31°C, the bees have a thermoregulatory problem: their body temperature enters the lethal zone for brood rearing.

Normal colonial cooling by *A. dorsata* occurs through a **fanning** and **"gobbetting" cycle.** Bees regurgitate dilute nectar or water, let it evaporate on their mouthparts (which cools down the water), and re-ingest the cooled droplet. That passes through the hot thoracic muscles, gathering heat, and is then regurgitated, gathering more heat to take to the outside. The wiggly section of the dorsal aorta as it passes through the petiole (Chapter II – 6) may be a heat exchange coil in the genus *Apis*. However, in extreme heat and still, humid conditions, evaporative cooling is curtailed. The question then becomes "how to become more efficient at dispatching excess heat?" The best way is to reduce mass and so cool more quickly. The bees, it seems, **defecate to become lighter.** The average "bee poop" from an individual worker of *A. dorsata* weighs about 25 mg, or 20% of its body weight. Calculations on the mass of the bees before and after defecation, coupled with the estimated heat capacity of bee faeces, suggests a 36% increase in a colony's cooling efficiency through its better use of water and fanning. That then calculates to a 15% increase in a colony's thermoregulatory efficiency.

CHAPTER V – 20

SWARMING & MOVING HOME

The **reproductive processes** of social insects can be thought of at two "levels". The **first is at the level of the individual**; by laying eggs, the queen reproduces by creating more individual bees (Chapters IV – 12 & VI - 23). The **second level is that of the colony**. Through reproductive swarming, one colony gives rise to two or more colonies. Of course, reproduction at both levels is interrelated, but it is only through the creation of sexual individuals that intergenerational reproduction occurs.

In bees, the reproductive process presents insights into the evolution and specialization in sociality.

- In **solitary bees**, each female makes her own nest and stores provisions for her larvae. Solitary bees constitute most of bee diversity (Chapter VI – 25). They are considered to be non-social but sometimes nest separately in aggregations.

- Some bees are **communal**. Several females of the same species and generation use the same nest, but each makes her own cells for housing her eggs. There is no separation of tasks, even though some co-operation between the bees occurs. Generally it is thought that communally nesting bees are closely related as sisters, or half-sisters.

- **Semisocial** bees live in small colonies of two to seven bees. In these colonies, the bees share some tasks in rearing brood, but only one female may be the egg-laying mother bee. Semisocial colonies are of sisters, or half-sisters.

- In **eusocial** bees, the members of the colonies clearly constitute distinct castes, the queen and workers being female and the drones being male. **Primitively eusocial** bees form temporary colonies. Once the colony produces sexual forms, its social structure deteriorates and all that remains of the colony are individual fertilized queens that initiate the next generation as a new colony. Almost all bumblebees, yellow jackets, and hornets follow that pattern, with winter as the period of interruption of social activity. In the tropics, such insects may be **truly eusocial**, like honeybees and stingless bees. They live in perennial, often large, colonies with definite castes in the social structure (Section IV). In some truly eusocial insects, there may be more than three castes: soldiers are recognized in ants and termites. Truly eusocial insects are characterized by having long-lived queens that periodically produce sexuals so that there is marked overlap of generations as colonies reproduce. The seasonal cycle in a colony of honeybees makes an excellent example of truly eusocial activity.

The social structure of a honeybee colony has been described according to the roles and activities of the three castes: queens, workers and drones (Sections IV & V). When a colony of honeybees is active in summer, it usually consists of one queen (the reproductive), some hundreds of drones (reproductives), and up to tens of thousands of adult workers (non-reproductive females)

Seasonal Cycle in the Colony

In general, "**colony**" can be defined to include several plant or animal organizations in which various individuals live together and interact in mutually advantageous ways. The social structure of a colony is a result of a long evolutionary history. Animals may live together (be social) in a number of ways.

Within the colony, co-operative brood care is taken on by the workers, and the queen lays the eggs. The drones are cared for by the workers and represent the colony's genetic investment, through male function, to the next generation. Overlap of generations is clearly part of the social structure, with the long-lived queen laying eggs from time to time to produce sons and daughters. The way in which the colony's investment into the next generation takes place involves not only reproduction of a new generation of sexually functional individuals, but also the providing of the new queen with all she needs to establish a new colony. It is by swarming that colony reproduction is accomplished.

Biologists believe that the ultimate reason for living is to reproduce. All the activities of a colony of eusocial insects, as for any solitary organism as well, are directed at reproduction, perpetuation of the species, and the passing of genes from one generation to the next. Natural selection favours combinations of genes (the genotype) that result in organisms (phenotype) with competitive traits. That idea, of **Darwinian fitness**, was introduced in Chapter V – 18 on foraging. How does the natural activity cycle of a colony of honeybees fit into that scheme of the ultimate goal of living?

Figure 20-1 presents the life cycle of a colony of honeybees in a temperate climate. The ideas can be transposed to other aspects of seasonality that dominate other parts of the world: e.g., rainy and dry seasons in the tropics and subtropics for the eusocial bees native to those parts of the world.

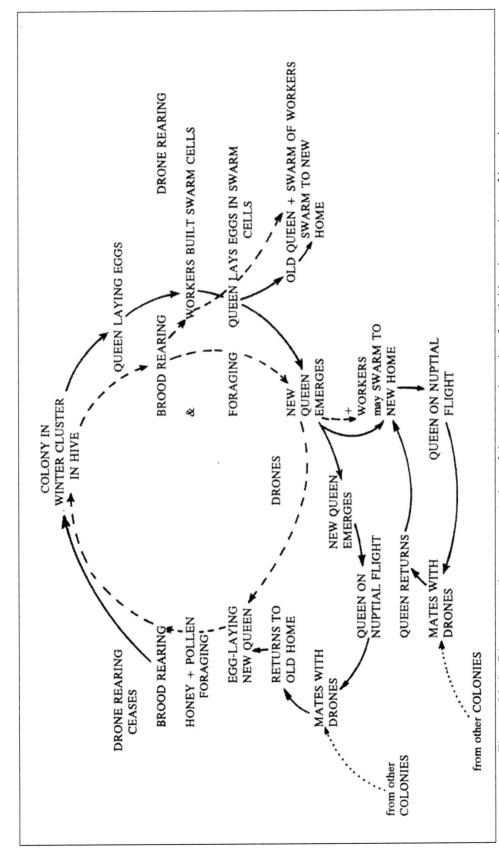

Figure 20-1. Diagrammatic representation of the annual cycle of activities in a colony of honeybees.

In **winter**, the colony forms its cluster over the resources stored from the previous autumn. The **winter cluster** maintains a temperature fluctuating around 25°C until the **queen starts to lay eggs**, sometime in **late January and early February**. Then the need for incubation requires that the colony boost its generation of heat so that the brood nest is maintained at a temperature of about 35°C (Chapter V – 19). The timing of onset of egg laying by the queen probably reflects that European honeybees have their origins in a climate much warmer in winter than that in Canada or the northern North America. On warm days in **February and March**, workers can fly from the hive for short, cleansing flights. They have been restricted to the hive for months, have had to feed, and have **recta** extended, sometimes to occupy most of the interior of the abdomen, with faeces (Chapter II – 5).

Once **spring** arrives, brood rearing accelerates. At this time, usually **late March**, the first flowers, such as willow and dandelion, start to bloom and provide forage. Prior to that, beekeepers may wish to help the colony recover from winter, and the depleted stores, by providing supplemental food as sugar syrup and pollen substitute. Once the flowering season is in full swing, drone rearing starts (**April**), and then queen rearing may start (**May**). By proper beekeeping management, queen rearing can be discouraged, an important step for maximizing honey production. If queens are reared in a colony, then **swarming** may ensue. On the other hand, workers may abandon queen rearing and destroy the queen cells they have built. The communication mechanisms involved (Chapter V – 16) are not understood. The swarming season is generally **May** and **June**. What is **swarming**? How is it **controlled**? And what are its **consequences**?

Swarming is the act of a family of bees leaving their home to establish a new home elsewhere. It involves the sudden departure of a proportion of the **adult worker bees** of a colony from its nest with a **queen** and sometimes also some drones.

Swarming occurs for a number of interrelated **reasons**: (1) too many bees, (2) too little space, (3) the hive has become too hot and humid through lack of adequately effective ventilation, (4) not enough storage space, (5) a failing queen, or (6) there is simply a genetic propensity to swarm. Reasons (1) to (4) all indicate crowding. A beekeeper can avoid that situation by proper management (Chapter VIII – 32). Swarming that results from a failing queen is probably initiated by the colony's raising several queens, and the reproductively effective swarms are those with virgin queens (that is a situation similar to secondary or after-swarming) (Figure 20-1). Africanized bees have a natural tendency to swarming and colony reproduction (Chapter VI – 24).

There is a specific series of steps that occurs over about a 2- to 3-week period when a colony prepares to, and produces a, swarm. To start, the colony seems to have an internal communication system by which increasingly crowded conditions are monitored (Chapter V – 16). The retinue surrounding the queen increases at this time, and she is fed more lavishly by the workers. The workers build queen cells and the laying queen increases her egg output by as much as 50%. Crowding becomes worse.

Then there is a switch. The queen is fed less, her ovaries and abdomen shrink in response, and egg-laying declines. Also, during this time, foraging declines: the hive is full! By this time, the immature queens, as pupae, are well on their way to adulthood (Chapter IV – 12). As the workers get ready to swarm, they gorge themselves on honey, and the excitement in the colony climaxes with the **buzzing run** dances (Chapter V – 17) (sometimes also referred to as whirring or breaking dances) over less than 15 minutes.

The swarm comprises at least a third to half the worker population of all ages (except the youngest callows), the queen, and often 50 or so drones. It leaves the nest in a spectacular cloud of buzzing activity as up to 10,000 bees lift into the sky at once and move together in one direction. The swarm may settle with a few to several hundred metres of the parent colony. A swarm that is settled tends to be calm: a dense cluster of bees hanging from a branch (Figure 20-2) or part of a building, or even lying on the ground. The mass of bees encompasses the queen. She maintains cohesion in the colony through pheromonal communication in flight and when settled. The workers use their Nasonov secretions to the same end (Chapter V – 16).

Normally, the first settling spot is not the new home for the colony. The swarm may remain in one place for a few hours to a few days before finding a new cavity (hive) in which to take up residence. Foragers scout for a new home and report back when they find it. They employ the foraging dance language to indicate direction of, distance to, and quality of prospective sites. The same processes of recruitment are used. Eventually, through that system a strong consensus is reached and the swarm moves off, following the same dance that initiated the swarming in the hive, to a new home. In European honeybees, that is usually within 1 km. In Africanized bees, swarming may be a multi-step process of temporary swarms (a process called **bivouacking**) over tens of kilometres (Chapter VI – 24).

Occasionally, scouts cannot find a suitable home nearby, and the swarm may make a home in the open. That is not a good situation and leads to the death of the colony, at the latest at the onset of winter. If all is well, the new colony (with the same, old, queen) builds comb quickly, and colony activities resume within a few days. The survival rates for such swarms are low, especially farther north, but if the colony survives the first winter, it is likely to become persistent in its new home for several years. If a beekeeper is fortunate enough to obtain and "hive" a swarm, it becomes part of his operation and continues to exist indefinitely. Diseases, especially parasitic mites, have all but extirpated natural, feral colonies (Chapter IX – 35).

Figure 20-2. Swarm of honeybees settled on a maple branch.

A single colony may produce more than one swarm. The first swarm with its mated queen is the **primary swarm**. Swarms that issue thereafter, in the same swarming period, are **secondary** or **after-swarms** and are led by a young, virgin queen. European races of honeybees produce only one or two swarms each year, but Africanized honeybees may produce as many as 6 to 12, with a number of those being primary swarms.

While the swarm is settling into its new hive, the old hive spends about a week (8 days) queenless before the new queens emerge. At this point, two things may happen. A new, virgin queen may initiate another swarm (an **after-swarm**). If two queens emerge at the same time, a battle between them ensues (the **battle royal**). If one new queen emerges before her sisters, she usually stings her sister, pupal, queens to death while they are helpless in their cocoons. The queen's sting is used for offence, and can be used several times: it is not barbed (Chapter I – 4). Once there is only a single queen left, she matures until ready for nuptial flights. The queen mates with 7 to 10 drones, over several days, while on several nuptial flights (Chapter VI – 21). Once mating is complete, she remains in the hive, laying eggs, until she dies or swarms. A healthy queen lays a staggering 1,000 to 1,500 eggs (almost half her own body mass!) per day when she is in peak production.

During the process of swarming, one can appreciate that the bees have prepared the old colony well for recovery from loss of so many workers. The old queen left the colony with abundant brood which—by the time the new queen has emerged, mated and started her reign—is active in preparing the brood chamber, and in foraging for stores to replace those removed during swarming.

As summer progresses, an active colony spends most of its efforts toward laying in stores of honey and pollen to last them through the winter. Strong nectar flows result in strong honey flows in the hive, and an abundance of honey. Late **June** to **August** is the period of greatest nectar flow and honey production. A fall flow in **September** is associated with the fall flowers: goldenrods and asters.

Drones are present in the hives throughout this time, even though the main mating season has passed. However, as the days draw in, the weather turns cool in **October**, and the last of the autumn flowers succumb to frost, the drones are dismissed from the colony and die outside. An overwintering colony rarely contains drones. Also, as winter approaches, the queen's egg production declines, and eventually ceases for a few weeks over the depths of winter. A colony requires between 15 and 30 kg of pollen and from 60 to 80 kg of honey annually to survive the winter in southern Canada.

The life cycle of the colony (Figure 20-1) can be followed graphically according to the production of swarms, colony strength, and honey production according to whether or not the colony is managed by a beekeeper (Figures 20-3 & 20-4). A beekeeper's main aim is not usually colony reproduction, unless breeding bees for queen, package, and "nuc" (nucleus colony) production is the plan. Most beekeepers want strong, populous colonies for honey, or other hive products, or for pollination. Swarming is not conducive to profitable beekeeping (Chapter VIII – 32 & Section X). It is an exciting process to watch and follow. Hobbyists may be interested to let nature unfold and swarming take place.

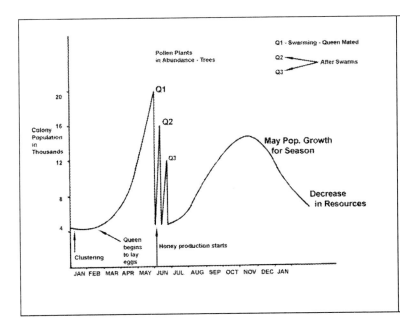

Figure 20-3. The annual cycle in an unmanaged/feral honeybee colony in southern Canada: Q1, Q2, and Q3 represent the issuing of a primary and two secondary swarms.

Swarms caught and placed in hives by beekeepers can be of great value, especially if collected in May or June. An old saying is that "a swarm in May is worth a load of hay, and swarm caught in June is worth a silver spoon, but a swarm in July is not worth a fly!" Late swarms have little chance of being able to establish, gather enough stores, and survive winter.

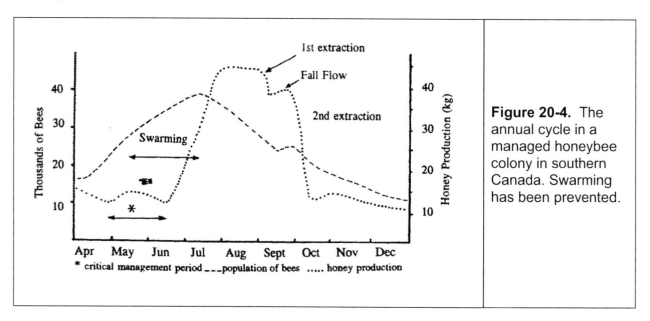

Figure 20-4. The annual cycle in a managed honeybee colony in southern Canada. Swarming has been prevented.

Four other processes in a honeybee colony can be related to reproduction: **supersedure**, **absconding**, **migration**, and **laying workers** (Table 20-1).

Supersedure is the replacement of one queen by another. It is requeening as done by the colony itself. The process is usually characterized by the old queen's becoming weak and her chemical influence over the colony failing. The workers produce **emergency** (or **supersedure**) **cells**, in the centre of a brood comb (Figure 20-5). They raise the larva as they would a queen, and the emergency cell resembles a normal queen (or swarm) cell, except for its position in the hive (see Table 20-1).

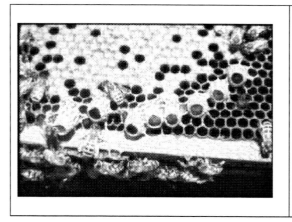

Figure 20-5. Emergency cells drawn from normal brood comb in preparation for supersedure.

Absconding is simply the abandonment of a hive by a colony. It is usually the result of poor conditions in the hive (heat or damp) (see also Chapter V – 19), or more often a dearth of resources. The colony recognizes that forage is in short supply in the foraging territory, and that it is likely to starve. Moving away is the best option for survival. In crowded apiaries of honeybees with small foraging ranges (e.g., *Apis cerana* in Asia [Chapters V – 17 & VI – 25]), absconding is common. It is blamed on the nature of the bees rather than on inappropriate management.

Migration is a form of absconding, except that it is cyclic and predicable. *Apis dorsata* is naturally migratory (Chapter VI – 25), as is *A. mellifera scutellata* in its native Africa (Chapter VI – 24). Both types of bee migrate out of areas where flowers are scarce, into areas where forage is more abundant. *Apis dorsata* migrates from the tropical rainforests on inland Malaysia once the rainy season starts and flowers are scarce. It moves to coastal areas where the rains are not as strong and flowers can be found. *Apis mellifera scutellata* migrates back and forth across the African savannah, following the abundance of flowering trees. The propensity for swarming by Africanized bees (Chapter VI – 24) has evolutionary roots.

The phenomenon of **laying workers** is well known to beekeepers, and is a situation that is undesirable. Workers start to lay eggs if a colony is hopelessly queenless (see Chapters V – 16 & III – 11). Eggs laid by workers are on the walls, instead of at the bottom of the comb. Several may be laid in one cell, and the pattern is often haphazard. All eventuating larvae can become only drones, so capped brood is characterized by raised caps. The activity is a final attempt to pass on genetic information, through male function, to the next generation of bees in the area. The colony is doomed! Even if a beekeeper intervenes, the colony will probably die.

Table 20-1. Comparison of Features and Characteristics of Swarming, Supersedure, Absconding and Migration, and Laying Workers

SWARMING	SUPERSEDURE	ABSCONDING	MIGRATION	LAYING WORKERS
Comparison of Features:				
1. True colony reproduction	No colony reproduction	No colony reproduction	No colony reproduction	No colony reproduction
2. New queen(s) reared	New queen(s) reared	No new queen	No new queen	No new queen, only drones
3. Colony strong/populous	Colony may be weak	Colony may be strong or weak	Colony may be strong or weak	Colony often weak and weakens
4. Seasonal May–July	Can occur anytime	Occurs during dearth, destruction	Occurs during dearth	Can occur anytime
5. 4-20 queen (swarm) cells	1-6 queen cells	0 queen cells	0 queen cells	0 queen cells, drone cells only (possibility of parthenogenically produced female eggs)
6. Queen cells near bottom of brood nest	Queen cells near centre of brood nest	N/A		N/A
7. Specialized cells not on surface of comb	Cells modified, brood cells on comb surface (Emergency cells)			
Caused by:				
8. Crowded hive, etc.	Weak queen or dead queen	Food shortage, heavy predation, fire, difficulty in regulating temperature, etc.	Annual cycles of availability of forage	Dead queen, requeening failed (Chapter VIII – 32)

SECTION VI

Biodiversity: Making More Bees & Being Different
Reproduction in the Colony, Swarming, Moving Home, Mating & Sex, Genetics & Breeding, & Diversity in Honeybees & Bees Worldwide

The idea of the "super-organism" (Section V) extends to **reproduction**. Although production of individual bees is indeed reproduction, and is the way that solitary bees continue from generation to generation, that does not apply fully to **advanced eusocial** insects, such as honeybees and stingless bees. In those insects, to reproduce, one colony must give rise to other colonies. **Swarming** is the process by which colonies reproduce. The colony raises new queens in special cells, called queen cups. Meanwhile, the parent queen, and a large number of workers, issue from a hive as a swarm. They set up a new colony in a new hive while the old colony back home continues. The process leading to swarming, and providing the parent colony with backup for the loss of population, is truly a coordinated colonial effort.

Changing queens does not require that the colony swarm. An aging, weakening, or dead queen stimulates the workers to raise a new queen from a fertilized egg somewhere in the brood. The process of **supersedure** assures the colony's persistence. Moving home, though, may not involve colony reproduction. During poor conditions, a colony of bees may just **abscond**. Other forms of reproduction in the hive may involve workers becoming egg-fertile and starting to lay. **Laying workers** are not inseminated, so their eggs become drones. The colony is doomed, but perhaps some of its genes are passed on to another generation by the mating of the drones with local queens.

As described in Chapter III – 11, the reproductive structures and **mating** rituals of honeybees are among the strangest known to science. Virgin queens must locate, and fly to, drone congregations. There, the drones compete to mate. For the few successful drones, death is assured in a blaze of exploding genitalia. The queen mates with several drones on several mating flights before she settles in as mother of the colony.

Genetics is not a familiar area for many people. Genetics in honeybees is complicated by the way **sex is determined**. Male bees (and males of Hymenoptera generally) develop from unfertilized eggs. They have only half the number of chromosomes as the females (workers and queens) that develop from fertilized eggs. The story is further complicated with variants of the genes (**alleles**) that are involved in sex determination. If a fertilized egg does not receive different variants, then it can give rise to a male bee. Thus, small populations of bees, in which individuals tend to be closely related, can and do produce sterile males from fertilized eggs. That is a conservation issue for wild bees, and of concern to honeybee breeders.

Selecting and breeding honeybees for desirable characters has proven worthwhile. It requires care in maintenance of stock, but has allowed beekeeping to overcome practical problems, especially in producing disease-resistant stocks.

Most of our knowledge about honeybees comes from the **Western honeybee**, *Apis mellifera*, and, within that species, **European races**. Nevertheless, there are other races of the Western honeybee, especially from Africa. They have different characteristics, and notoriously have given rise to the so-called **Africanized**, or **killer**, or **assassin honeybee**. Beyond the range of the Western honeybee live the Asiatic species. Some, such as *A. cerana*, live in cavities and are kept by beekeepers. Others live in exposed, single-comb colonies, e.g., *A. dorsata*, *A. laboriosa*, and *A. florea*, and are exploited for their honey and wax.

The diversity of bees is estimated at **30,000 to 40,000 species** worldwide. They are classified into a number of families, each with distinctive traits. The family, Apidae, is the most diverse and includes a large number of species of solitary bees, and relatively few species of **eusocial** species (with castes), such as **bumblebees**, **honeybees**, and **stingless bees**.

CHAPTER VI – 21

MATING

Mating is the activity by which sexual union between individuals takes place and includes any reproduction involving two sexes. The sexual union involves the transmission of genetic information, contained in **gametes**, from the donor (male) to the recipient (female). In most plants and many animals, individuals are **hermaphroditic** (each individual is both male and female), so that male function is the donation of gametes (**sperm**) and female function is the production of female gametes (**eggs**) and receipt of male gametes. New individuals result from the union of a sperm cell with an egg cell (**fertilization**) in most plants and animals (Chapter IV – 12), but in some, new individuals are produced without the union of gametes. That is called **parthenogenesis**. In honeybees, and Hymenoptera generally, eggs that develop into female individuals do so after the union of sperm and egg nuclei in the egg, but the males (drones) develop parthenogenetically from unfertilized eggs. That sort of reproductive process is called **arrhenotoky** or **haplo-diploidy**. Haplo-diploidy refers to the number of chromosomes in cells. In the gametes of most animals and plants, there is half the number of chromosomes (haploid condition) as in the parent organism (diploid condition).

Honeybees use **polyandry** (multiple mating), a system in which females have several mating partners. Thus, the female offspring (workers and queens) from a single queen honeybee may have several different fathers. The feminine part of the family in a honeybee colony consists of **sisters** and **half-sisters**. In general, the mates of an individual queen are drones from colonies not her own. Brother–sister matings are uncommon in nature and carry high risks through the genetic consequences of inbreeding (Chapters VI – 22 & 23). In honeybees, the result is production of fertilized eggs that would become diploid drones if the workers did not detect and destroy them. In other rare and endangered bees, small populations increase the likelihood of brother–sister matings and production of diploid drones, and those are mostly sexually sterile. This is now cause for concern in conservation of pollinators.

Mating in honeybees is best understood in the Western species (*Apis mellifera*). After a queen has eclosed from her cell, it takes her 5–10 days to mature to readiness to mate. She departs the hive at the urging of the workers (Chapter V – 17), at first on **orientation flights**. Once somewhat familiar with her surroundings, she embarks on mating flights to the **drone congregation areas** where mating (copulation) occurs. Drones, too, must mature for a few days after they eclose. At 6 to 8 days old, adult drones first fly to congregation areas. These may be a kilometre or more from their home hives. They are often associated with particular landmarks, such as the edge of a woodlot, and may be anywhere from 3 to 40 metres above the ground. The drones are really ready to mate once they reach about 12–14 days old.

The weather for mating must be balmy, suitable for a picnic. In general, the air temperature should be 20°C or above, winds should be light at less than 5 metres/second, and it should be sunny or partly cloudy. In southern Canada, June is the perfect month for honeybee mating. Mating usually takes place from 2–4 p.m.

As the queen enters the drone congregation area, the drones sense her presence by vision (large eyes: Chapter I – 2) and smell (long, highly sensitive antennae [Chapters I – 2 & VII – 27]; queen substances [Chapter V – 16]). The queen is attracted by the drone pheromone (Chapter V – 16) released into the air at a congregation site. A crowd of drones chase the queen and by so doing, form a "**drone comet**".

The actual copulation between drone and queen has earned the honeybee the dubious title of one of the most bizarre mating activities in the animal kingdom. The first drone to reach the queen mounts her; then, in copulation, his genitalia actually explode as he is flipped onto his back, paralyzed! Attached to the queen for a few seconds, he is dragged through the air upside down. This process is fatal to the drone. He falls to the ground, leaving his genitalia anchored in the queen's vagina, along with parts of his internal organs. Those remains are termed the "**mating sign**". A second, third, and fourth drone may mount the queen during one nuptial flight. Each must remove the previous drone's "mating sign". Perhaps the strange and elaborate appendages of the drone's penis (endophallus) are used for that purpose. At the end of the nuptial period on a given day, the queen returns to her colony. There the workers remove the mating sign of the final copulation of the day. This ritual goes on for some time, over several days (days off for bad weather) with up to four nuptial flights so that the queen has mated with, and carries the sperm of, 7 to 17 drones for the rest of her productive life.

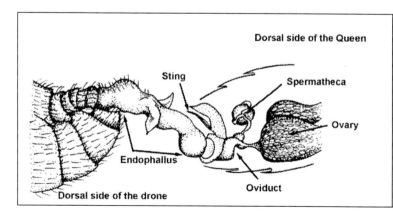

Figure 21-1. Semi-diagrammatic illustration of the juxtaposition of the genitalia of a drone and queen during copulation in the honeybee.

Chapter 21
Mating

The mating time in different species of honeybees takes place at different times of day. In places where there are several species of honeybee, the temporal isolation provides a way for the species to remain true during mating. The structures of the drones' endophalli differ between species. Timing and anatomy both contribute to reproductive isolation (Chapter VI – 25).

The **semen** of the drones is retained and moved within the queen by contractions of the muscles of the common oviduct (Chapter III – 11) and reaches the spermatheca within about 40 hrs. The queen starts laying 2 to 4 days after her final mating.

CHAPTER VI – 22

BEING FEMALE OR MALE: SEX DETERMINATION

Chapter VI – 21 explains some aspects of sex determination in Hymenoptera, and in bees particularly. The **haplo-diploid** system is briefly introduced as a prelude to **mating**, but more genetic explanation is needed to understand how sex is determined in this strange system.

The cells of the body divide as growth and maturation progress. The genetic information in all the body cells is identical, even though the cells making up a given tissue can be strikingly different (e.g., blood cells versus nerve cells). The genetic information for each cell is carried on the chromosomes. Most organisms have a set number of paired chromosomes (the **diploid number**). As the cells divide, the chromosomes are duplicated and the number is retained through the process of mitosis. However, reproductive cells (gametes) unite, then divide as a new organism develops. Gametes are produced through the process of meiosis (reduction division) so that the cells (sperms or eggs) have only half the number of chromosomes, one of each pair (the **haploid number**).

Sex is determined in most animals by **sex chromosomes**. The sex chromosomes typically are designated **X** and **Y**. For example, in mammals including ourselves, females have paired X chromosomes **(XX = ♀)**, but males have one X and one Y (i.e., in males the sex chromosomes are mismatched in terms of the idea of being a more or less identical pair) **(XY = ♂)**. Because male mammals produce two kinds of gametes, they are called the heterogametic sex, and the females the homogametic sex. In birds the situation is reversed.

men : X_{sperm} Y_{sperm}
women: X_{eggs} X_{eggs}

After fertilization, which is the fusion of the gametes, the diploid condition is restored in the **zygote**. Random union by either an X or a Y chromosome from the male mammal with the X eggs of the female results in a 50:50 chance of the zygote's being male versus female.

XX or **XY**
girl boy

In most insects, sex is determined as described above, but not in the case of Hymenoptera. Female honeybees—workers and queens—arise from fertilized eggs,

161

but drones arise parthenogenetically from unfertilized eggs. The characteristic diploid number of chromosomes for all species of honeybee is 32 (2 sets of 16 in pairs) in females, but males are haploid with only 16 chromosomes (no pairs). Drones do produce sperm which, like the body cells, are haploid. During the process of spermatogenesis (making sperm), meiosis probably stops at the point at which the otherwise paired chromosomes would separate (i.e., as they do in females) . In the process of fertilization, each parent contributes its genetic material to female offspring, but drones have genetic information only from their mother queen. **Arrhenotoky** or **haplo-diploidy** is the genetic mechanism of sex determination in Hymenoptera. The result in honeybees is that drones have only daughters, no sons! And drones have no father, but do have grandfathers!

One of the questions that has arisen is "Is the haploidy required for hymenopteran maleness?"

In the 1940s, Whiting unraveled the sex determination in the small parasitic wasp, *Bracon hebetor* (*Habrobracon junglandis*). His findings can be applied to the honeybee, and the following explains briefly his experiments. In essence, he inbred his wasps and determined what sexes would eventuate and what their genetic constitution would be.

Whiting inbred the offspring of the first generation (F_1) so that among the genetic recombinations represented in the second generation (F_2) would be individuals with two sets of identical chromosomes (i.e., a condition termed homozygosity). White determined that, although those wasps were diploid, they were male. In the simplest cases, sex determination was found to be under the control of a single gene, and the individuals that are heterozygous (having differences in the genes present on each of a pair of chromosomes) for the sex-determining gene (or allele, the actual nature of the gene) always developed into females. Homozygous individuals, those having two identical forms of the gene (i.e., identical alleles) in this locus, developed into diploid, biparental males.

Figure 22-1 describes the experiment and the process. The haploid drone can carry only one allele (for example call it "c") in the sex-determining locus. The female is diploid with two different alleles (for example, "a" and "b") for this sex-determining gene. The gametes—egg genotypes—occur in two kinds: "a eggs" and "b eggs". The sperm genotype can be only "c". The mating is tri-allelic. In the F_1 (first filial), the expected offspring results: after fertilization, XaXc and XbXc are diploid females; without fertilization, the haploids (hemizygotic offspring) are Xa males and Xb males.

Next, a haploid Xa male is mated with an XaXc female (a di-allelic mating). In the ensuing generation (F_2), a quarter of the progeny are homozygous diploid drones and the other three quarters are heterozygous females.

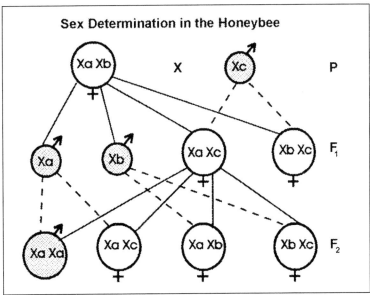

Figure 22-1. Sex determination in the Hymenoptera (e.g., honeybees). Females (♀) are diploids and are heterozygous at the sex locus X. Males are hemizygous (haploid) (♂, small circles) or are homozygous (♂, large circles). In F_1, a tri-allelic parental mating was experimentally arranged by instrumental insemination. In the F_2, the di-allelic mating, again by instrumental insemination, results in half the progeny being homozygous diploid drones and the other half heterozygous females. Mating with both drones results in 1/4 diploid drones and 3/4 females.

Woyke (1996) developed a technique to raise diploid drone honeybees to maturity. Normally, in the hive, these bees are detected by workers and killed as young larvae, but if incubated separately they can be raised to maturity. Woyke and his co-workers also studied spermatogenesis in diploid drones and found they produced diploid spermatozoa. Diploid sperm would result in triploid zygotes following fertilization, if that were even possible. The genetic consequences of such recombinations would probably be inviable.

From that sort of research in honeybees, and the results of elaborate series of genetic experiments, it is thought that the gene for sex determination has from 6 to 18 forms (alleles). As long as the alleles are different in the two chromosomes of the diploid offspring, it is female. In haploid offspring, there is only one gene and no matter what allele is present, the offspring are male. The implications from this sort of research are important in bee breeding and selection (Chapter VI – 23). Clearly, a bee breeder must maintain genetic diversity in the bee mating yards if healthy, strong colonies are to be expected from progeny. It has been suggested that in some places, beekeeping has failed because of small populations and the effects of inbreeding, production of diploid drones, and dwindling of the population of bees in their colonies.

Implications for the problem of conservation of pollinators are also of concern (Chapter VI – 21).

The issues of sex determination are more than matters of academic curiosity. Studies in population genetics of honeybees and other bees are necessary to establish the expected viability of brood in small populations. When populations are small and well isolated, inbreeding usually occurs. High levels of homozygosity increases the likelihood that brood will be destroyed by the workers, or that diploid males would occur in other bees. If the number of sex alleles is larger in the population, the chance of homozygosity is lower, and so there is a much greater percentage of survival (Figure 22-2).

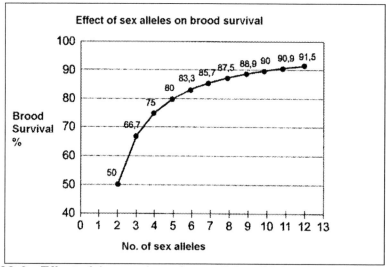

Figure 22-2. Effect of the number of sex alleles (N) in a honeybee population on the average percentage survival of brood in bee colonies.
Survival rate S% =100 (N–1) / N.

In the **Cape honeybee** (*Apis mellifera capensis*) (Chapter VI – 24) of South Africa, females may arise from unfertilized eggs. In this form of parthenogenesis, called **thelytoky,** recombinations of chromosomes occur automatically after meiosis, and those restore the diploid number of chromosomes and somehow maintain *heterozygous alleles at the sex locus*, the necessary condition for female development. The exact mechanisms are subject to ongoing research. Because laying workers (Chapter V – 20) of the Cape honeybee can lay eggs that develop into more workers, and because they have a propensity to invade colonies of other races of honeybees, e.g., *A. m. scutellata*, they are regarded as a pest of beekeepers. They have been expanding their range in southern Africa and can be regarded as "invasive".

Sometimes, but rarely, sex determination goes wrong and **sexual mosaics** or **gynandromorphs** result. These are individuals that are part drone and part worker. Various forms have been found: bilateral with one side male and the other female, segmental with one end male and the other female, and some with mosaic patches of different sexes throughout the body. Some may have functional female or male sex organs. The primary mechanism leading to gynandromorphy is **polyspermy,** which is the result of more than one sperm penetrating the egg, with cleavage and subsequent development of an accessory sperm nucleus. The haploid sperm nucleus leads to production of male tissue, while the zygotic nucleus (formed by the union of the egg and primary sperm) produces the diploid female tissue.

CHAPTER VI – 23

GENETICS & BREEDING

Although a number of terms from genetics have been used in Chapters I – 1 and in Chapters VI – 21 & VI – 22 on mating and sex determination, it is worth presenting them in a slightly more formal way.

Genetics is the science of heredity that studies the structure and function of genes and the way genes are passed from one generation to the next.

A **gene** is the determinant of a characteristic of an organism. Genes are items, segments, of chemical information coded in the DNA. As science comes to understand in greater detail the nature of genes, they are recognized as entities that are not as simple as once was thought. Genes are responsible for species and **variation** between individuals within a species, and for the differences between species. The more organisms differ from each other, the greater are the differences in their genes.

The **genome** is the total complement of genes carried by an individual or by a species. It is sometimes equated to the set of chromosomes that an organism has. In eukaryotes, that can be defined as the amount of genetic material in the **haploid** set of chromosomes. The honeybee's genome is 16 chromosomes (N = 16), in human beings it is N = 23. More recently, the term genome has assumed more of the former meaning (number of genes), as the total genomes for human beings and honeybees have been elucidated.

Genomics is the study of the structure and organization of entire genomes. The genome includes sequences of chemical codes that specify genetic information that, in turn, translates into the characteristics of the organism itself. It is the **genotype** (the genetic make-up) of an organism that predicates the **phenotype** (the suite of characteristics) of that organism.

Chromosomes are complex molecules consisting of DNA (deoxyribonucleic acid) that is composed of orderly sequences of base pairs of nucleotides (guanine and cytosine, adenosine and thymine, and uracil instead of thymine in RNA). The strings of nucleotides are complexed with protein. The arrangements of long sequences of base pairs comprise genes. They are the chemical code by which the organism makes so many different kinds of proteins. The arrangements often contain variations so that similar genes may code for different proteins and phenotypic variability may result. The genotypic variants are **alleles**.

An **allele** is one of two or more alternative forms of a single gene. A genetic **locus** is the place on the chromosome where a particular gene and its alleles reside.

Mostly, the physical location i.e. locus for a gene is not known, but it is understood that the alleles for a given gene are at the same place on a given chromosome. The recently completed **"Honeybee Genome Project"** allows for that to change quickly. That huge international project to sequence the entire genome of the honeybee was completed in 2006. Thus, honeybees join the exclusive club that includes human beings, chimpanzees and dogs for which the entire genetic sequence has been elucidated. Why would over $7.5 million be spent on such a project?

From what has been presented so-far in this book, the marvels of how such tiny animals, with so few cells, can perform such fantastic tasks, form lasting memories, and integrate them into social harmony has fascinated you, the reader, and the whole scientific world. Understanding how genetic coding controls such complexity has immense practical value for life sciences, human biology, and, of course apiculture.

It is suggested that the honeybee has about 10,000 genes, compared to 25,000 in human beings. Already, aspects of the genetic control of bee behaviour have been elucidated, and genes associated with diseases and resistance to them used for investigation into Colony Collapse Disorder (Chapter IX – 34).

Any change in the genetic material not caused by genetic recombination (see below) is called **mutation**. Mutations can be detected molecularly by comparing the genetic codes of mutant life forms with what are considered normal. In the simplest case there are two forms of a gene. For example: in humans the dominant allele **R** codes the ability to roll one's tongue into a U shape. If both alleles in one locus are the same, the individual is said to be **homozygous** for that gene (**RR**; it is called a homozygous dominant trait). If the individual possesses one gene for "no ability to roll the tongue", i.e., the **r recessive allele** and the **R dominant allele**, the individual is **heterozygous** for that trait (**Rr**). The person still can roll the tongue into a U shape. Only when the two recessive alleles are present–the **rr** allelic combination—an individual cannot roll his/her tongue; therefore, the recessive trait is expressed. Two genotypes represent the phenotype of the ability to roll their tongues, and one genotype does not and corresponds to the phenotype of lacking the ability to roll the tongue.

The ABO blood group system in humans is a well-known example for **multiple alleles** (I^A, I^B, and I^O), although only two of the alleles can be present in one individual (remember, there are only two chromosomes in a pair). In plants, apples have multiple alleles for incompatibility between pollen and stigma, which is why pollinizer trees (pollen donors) are needed in orchards with the main fruit-bearing variety (Chapter X – 45).

Genetic recombination is a process by which parents with different genetic characters give rise to progeny so that genes which differ in each of the parents are associated in new combinations.

Two types of chromosomes are present in most higher animals and some plants: the **autosomes** and the **sex chromosomes**. As explained in Chapter VI – 22, the situation in Hymenoptera is different: sex is not determined by sex chromosomes.

The growth and maintenance of an organism is achieved through cellular growth and division. The cells multiply by division! When body cells grow and divide, the genetic material also divides in the process of **mitosis**. However, when gametes (sperms and eggs) are made, they must contain half the number of chromosomes as the body cells so that when they unite in fertilization, the diploid number of chromosomes is restored. **Meiosis** is the process of reduction-division by which gametes form. Meiosis is how the queen honeybee produces haploid eggs, but the process of meiosis does not occur in drones. They already have only half of the complement of chromosomes ("Eric The Half A Bee", for Monty Python fans!)

The genetic organization of a honeybee colony comprises a super-family of sisters and half-sisters in the worker force, sisters and half-sister queens when reproductive females are produced, and genetically identical brothers. Genetic differences between half-sisters arise because the queen mates with several drones, and each sister family has the same mother and father, but the half-sisters differ by paternity. Sisters also have genetic differences, but those arise because of the different ways in which the same sets of genes combine during the process of reduction division to eventually result in eggs. There is much less opportunity for recombinations to arise when sperm are being produced, but they do.

Calculating how closely honeybees in a colony are related to one another is interesting. The sister families have an average of 75% of their genes in common, so they are **super-sisters** (queen mother and the same drone father). The **half-sisters** share 25% of genes in common (if drone fathers are unrelated). Full-sister relationships exist if "brother " drones inseminate one queen.

One can appreciate why there is a question about the possibility that super-sister workers should preferentially rear larval super-sisters rather than larval half-sisters. But do they? The answer for researchers is not clear (Chapter V – 16).

With the array of relatednesses that honeybees present—super-sister queens and workers, full-sister queens and workers, half-sister queens and workers, and genetically defined drones—bee breeding has taken a number of approaches to improving stock and understanding selection. The potential for obtaining inbred lines, hybrid offspring, selected stock, and line breeding to improve honeybees is great, especially through instrumental insemination. Nevertheless, the ever-present potential problems of inbreeding depression and diploid drone production mean that bee breeders must be careful.

Figure 23-1. A queen honeybee being instrumentally inseminated. The queen has been anaesthetized with CO_2 and is held in the restraining tube. The hook holds her vagina open while semen taken from a drone is injected.

Inbred-hybrid breeding uses the potential for superior traits to surface in the phenotype when two inbred lines are crossed. Some years ago, Starline (a hybrid with *A. m. ligustica* [Italian] origins) and Midnite with *A. m. caucasica* [Caucasian] origins) lines of honeybees were bred this way and were well used. The selection and breeding progress through **recurrent selection**, by which the progeny from repeated hybridizations may be discarded or maintained and, if maintained, may be marketed. The famous Buckfast line of superior honeybees was developed this way.

Strict inbreeding can be achieved by various means. Even self-fertilization can be achieved by using mother–son matings under highly artificial conditions. The equivalent of mother–daughter matings is achieved by mating a virgin queen with one of her brothers (her brother's genome is essentially fully contained within the his vergin sister's). The genetic equivalent of sister-sister matings can likewise be arranged. The problem arises because of the high numbers of diploid drones that are produced (Chapter VI – 22). Nevertheless, for investigating how characters are inherited, and for making rapid selections, the techniques have value.

Closed population breeding allows bee breeders to progressively improve stock by using queens and drones from colonies that show desirable traits. The bees may be allowed to mate in isolated mating yards, away from a source of drones of unknown origin, or breeding may be done by instrumental insemination. Closed population breeding has produced various honeybee stocks with interesting and superior traits.

Both high and low pollen-hoarding strains have been bred. Improvements in honey production have been achieved. High affinity to collecting pollen of alfalfa has been bred into honeybees with a view to overcoming some of the problems with alfalfa pollination (Chapter X – 45). In Ontario, lines of honeybees with resistance to tracheal mites and *Varroa* mites have resulted from careful breeding regimes (Chapter IX – 35).

Through **simple genetic studies**, several genetic traits have been elucidated as to their modes of inheritance. Because drones are haploid, recessive alleles are often expressed without their presence being masked by dominant alleles. For example, several alleles have been discovered for colours of compound eyes, for colour of pile, for hairlessness in honeybees, and for subtle chemical differences in some enzymes.

More **complex, heritable traits** require longer term studies involving controlled matings. One of the first demonstrations of the effectiveness of the method was selection for the number of hamuli; a high line with about 27 hooks and a low line of about 14–15 hooks resulted after 10 generations. More practical was the selection program for resistance to American Foul Brood (Chapter IX – 34). Using experimental inoculations with causative bacteria-selected colonies, the incidence of the disease dropped from about 70% to about 20% over 2 or 3 years, and was maintained at that low level by stock maintenance for another 12 years. Selective breeding, over only three generations, has been used to obtain lines with affinity to alfalfa pollen collection. Similarly, in three generations, selective breeding resulted in a fast line and a slow line of honeybees with respect to their speed of collection of sugar syrup. Selections over about 3 years resulted in honeybee stocks that were resistant to tracheal mites in Ontario and New York.

Hygienic behaviour is also heritable and has been found to be a highly desirable trait in resistance to diseases and mites. Hygienic bees are quick to uncap and remove the dead or damaged brood. There seem to be two heritable components to a colony's hygienic behaviour: uncapping and then removal of corpses. Those traits have been selected in attempts to breed stocks that are resistant to brood diseases, such as American Foul Brood (above) (Chapter IX – 34) and *Varroa* mites (Chapters IX – 34 & 35). Grooming behaviour can also be regarded as a kind of hygienic activity. It appears that workers of *A. cerana* groom each other for the ectoparasitic mites, *Varroa*, killing them in the process. *Apis mellifera* has less of a tendency for this behaviour, but it can be considered as a trait for breeding.

Defensive behaviour is also known to be heritable. Colonies of honeybees vary in their predisposition to defence (e.g., especially Africanized bees) and the release of alarm pheromone. Nevertheless, breeding programmes are advocated to facilitate beekeeping and to reduce the public's risk of attack by especially defensive honeybees.

With all **selective breeding programs**, if the superior performing stock is to be maintained, beekeepers and bee breeders must keep strict control on their program. They need to control variability in breeder hives, and to keep it high enough to avoid inbreeding problems. The more mates a queen takes, the greater is the genetic variability of the progeny. However, they must control the variety of matings through knowledge of the queen's genomes, and of the drones' genomes. That way, controlled genetic recombinations can be achieved in line breeding in isolation yards and by instrumental insemination. Just as with any superior stock used in agriculture, whether animal or plant, maintenance is crucial and expensive (Chapter X – 44).

Little research has been done on the genetics of other bees. A little line breeding has been tried with the **Asiatic hive bee**, *Apis cerana*, with such traits as honey hoarding, gentle disposition, and less tendency to abscond reported. The genetic potential for this manageable honeybee remains unexplored and untapped. **Alfalfa leafcutter bees**, *Magachile rotundata*, are extensively used in crops. In Western Canada with its short summer season, losses of those bees occurred because some of the population in culture had a tendency to try to produce two generations per year. Breeding has achieved stock that is **univoltine**, i.e., it goes through only one generation each year because the progeny enters its winter quiescent stage after maturing almost to the pupal stage. Variability in the **blue orchard bee**, *Osmia lignaria*, has also been recorded as having potential for selection. Similarly, genetic variability is known for various species of **bumblebees**.

Chapter VI – 24

HONEYBEE RACES

Like all animals and plants, honeybees are variable. Within a colony there is generally less variability than between colonies, and between geographical localities, one expects to find even more variability. The variability in form, colour, hairiness, and behaviour are shown by the phenotype, which is the manifestation of the effects of the genotype. Within the species of the Western honeybee, *Apis mellifera*, **intraspecific variation** is well known. That variation allows for selection of traits of interest to beekeeping (Chapter VI – 23) because matings within a species result in viable and reproductive offspring. Interspecific matings (between members of different species) almost always result in unfertilized eggs, death of embryos, inviable young, or sexually sterile adults. (Reproductive isolation is discussed briefly in Chapters III – 11 & VI – 25). **Races** and **subspecies** (terms that have almost the same meaning) are categories of interspecific variation. A population that can be distinguished from other populations of the same species by several genetic characteristics can be described as a subspecies, race, or breed. Breeds, or lines, of honeybees are mentioned in Chapter VI – 23. Races are mentioned in passing in various Chapters (III – 11, V – 17 & 20).

It has been suggested that the Western honeybee made its way into Asia Minor, Europe and Africa in several waves of immigration form Asia about 1 million years ago. The northern wave is characterized by the so-called German race of honeybees, *Apis mellifera mellifera*. The African races characterize the southern wave. The middle wave is characterized by the races of southern Europe (Figure 24-1). Another suggestion is that the evolutionary split in the ancestry of *A. cerana* and *A. mellifera* occurred 6 to 8 million years ago and that the racial diversity in the Western honeybee resulted from two range expansions from Africa into Europe and Eurasia about 1 million years ago.

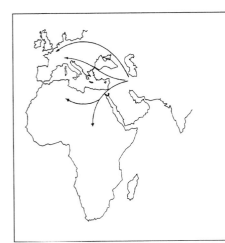

Figure 24-1. Suggested waves of the invasion of Europe and Africa by *Apis mellifera*. The northern wave comprises *A. m. mellifera*; the central wave *A. m. caucasica, carnica,* and *ligustica*; and the southern wave *A. m. intermissa, monticola,* and *scutellata*. Another suggestion is that *A. mellifera* evolved in Africa and then invaded Europe and western Asia

The races of *Apis mellifera* and of other *Apis* spp. can be distinguished by various **external characters**, such as:

- **Colour**—the first dorsal segments of the abdomen vary between light yellow and entirely dark (Figure 24-4);
- **Size**—there are small and large races;
- **Size of the body parts**—length of tongue (proboscis), hind leg, hind femur, forewing, and plates on the tergum vary consistently between races;
- **Hairs**—specific differences in the length of the cover hairs (pile) on the upper side of the abdomen can be diagnostic features (Figure 24-2);
- **Veins of the wings**—the wing venation is used to calculate the cubical index of the forewing (Figure 24-2), a feature used to distinguish Africanized bees from European races;
- other **biometrical** characters can also be used, e.g., number of hamuli (Figure 24-2), width of the hind metatarsus.

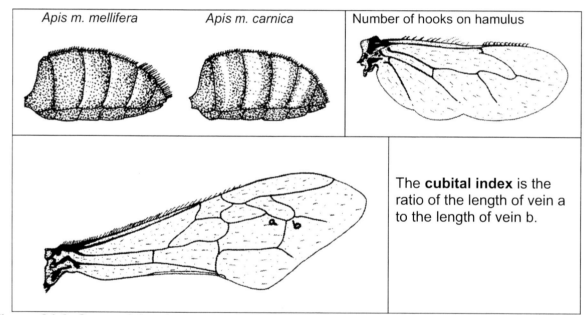

Figure 24-2. Some of the characteristics used to distinguish between races of honeybees. The tomentum, or pile, on the abdomen differs between races (0.5 mm long in *Apis m. mellifera*, as compared to 0.3 mm in other races), the number of hooks on hamuli varies, and the forewings carry important characters, such as total length (<9.01 mm in Africanized bees, longer in European races) and the cubital index (which ranges from about 1.7 to 2.7 according to race).

Molecular diagnoses can also be used:

- **Mitochondrial DNA (mtDNA)** can be extracted from the genetic and protein-synthesizing system of all life forms. Variations in the sequences in which the base pairs are arranged (Chapter VI – 23) can be used to calculate how closely related, and how different, are populations, species, and higher taxonomic divisions.

- Several forms of various enzymes, called **isozymes**, are commonly found in populations, each coded by a different allele of the encoding gene. The frequency with which the different isozymes occur can be characteristic of a particular population, breed, race, or subspecies.
- The **chemical nature of pheromones** varies between species and races of honeybees.

Within the species of Western honeybee, at least 24 races have been recognized. Dr. Franz Ruttner published his seminal work "Biogeography and Taxonomy of Honeybees" in 1987 (Table 24-1 and Figure 24-3).

Table 24-1. Geographically Recognized Races of the Western Honeybee, *Apis mellifera*, Categorized into Four Groups: African; Near Eastern (from Asia Minor); Central Mediterranean and Southeastern European; and Western Mediterranean and Northwestern European. *A. m. pomonella* (noted in parentheses) lives in the Tien Shan Mountains of Central Asia, 2000 km east of the previously estimated range for the Western honeybee.

African	**Near Eastern**	**Central Med.** **SE European**	**Western Med.** **NW European**
A. m. scutellata *A. m. adansonii* *A. m. litorea* *A. m. monticola* *A. m. lamarckii* *A. m. capensis* *A. m. unicolor* *A. m. yemenitica*	*A. m. anatoliaca* *A. m. adami* *A. m. cypria* *A. m. syriaca* *A. m. caucasica* *A. m. meda* *A. m. armeniaca* *(A. m. pomonella)*	*A. m. sicula* *A. m. ligustica* *A. m. carnica* *A. m. macedonica* *A. m. cecropia*	*A. m. sahariensis* *A. m. intermissa* *A. m. iberica* *A. m. mellifera*

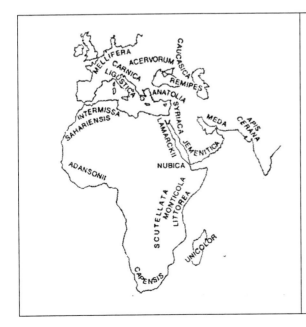

Figure 24-3. Approximate locations of some races of the Western honeybee (*Apis mellifera*) in Asia Minor, Europe and Africa.

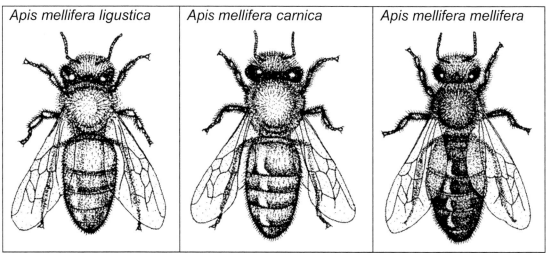

Figure 24-4. Workers of three of the important races of Western honeybee. The Italian bee (*A. m. ligustica*) is the smallest and lightest in colour, while the German bee (*A. m. mellifera*) is the darkest. The Carniolan bee (*A. m. carnica*) has a greyish appearance.

The Western honeybee is now found throughout the world. From its natural range in Europe, it was taken to North America with settlers to the Thirteen Colonies in about 1620. It was transported to the Spanish and Portuguese colonies at about the same time. It was not imported into Canada until probably about 200 years later. It was not until 1822 and 1842, respectively, that it was taken to Australia and New Zealand. Honeybees made their way westward in North America with the pioneers. At about that same time, colonies were transported to Japan and China (Figure 24-5).

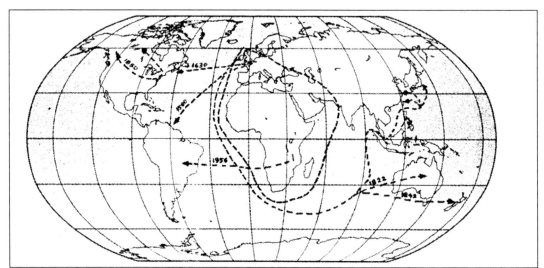

Figure 24-5. The natural range of the Western honeybee in Europe, Africa, and the Near East, with arrows indicating intercontinental transport from Europe, and then in 1956 from Southeastern Africa.

Although there has been much international traffic of Western honeybees, perhaps the most notorious recent event was the transport of the African race, *A. m. scutellata* to Brazil in 1956. Those bees were taken there as part of a breeding program to develop a tropically adapted race of productive honeybees. It was hoped that a line of superior bees for honey production would eventuate. The result, after the accidental release of the African stock, accompanied by their swarming, was the rapid colonization of most of South and then Central America with the so-called "killer bees", "assassin bees", or Africanized bees (Figure 24-6).

Although the Africanized bees became established in parts of the Southwestern USA in 1995, they have been slow to migrate eastward around the Gulf states. The suggested extent of their natural range expansion, especially to the north, has been hotly debated. The Africanized bees do not seem well adapted to clustering in cold weather (Chapter V – 19), so it is thought that they will be unable to establish in regions with cool and cold winters. The major problems with these bees are (1) that they are a public nuisance because of their highly defensive behaviour—an individual sting is no more potent that that of any other Western honeybee, but they sting en masse; (2) that they are difficult to manage and therefore require that the beekeepers wear protective clothing, which can be uncomfortable in hot weather; (3) that they have a propensity to swarm; and (4) that they may produce reduced honey yields.

Figure 24-6. The spread of the Africanized bee (a hybrid between the European race being used in Brazil prior to 1957 and *Apis mellifera scutellata* imported from Southeast Africa for a breeding program in 1956). The latter escaped from confinement in 1957. The rate of spread is given by annual increments in range until 1968; 3 years later, the bees crossed the Amazon. By 1975 and 1982 they were in Venezuela and Central America. By 1990, they were in Mexico and had crossed into the USA.

The features of the various races of honeybee reflect natural circumstances in their natural ranges. The propensity of *A. m. caucasica* to produce lots of propolis (Chapters V – 19 & X – 42) is probably an adaptation to living in a region with cold and snowy winters, the Caucasian mountains. The acknowledged length of the proboscis, greater than in other races, has resulted in their being advocated for pollination of red clover (Chapter X – 45) with its long, tubular florets and deeply hidden nectar. Perhaps the gentle disposition of the main European races has come about because of selection by beekeeping for easily managed bees in places where populations of natural predators have been reduced by human influence. The propensity for swarming in Africanized bees may have its evolutionary roots in the migratory behaviour of the ancestral African race (Chapter V – 20).

Table 24-2 summarizes some of what is known about the characteristics of various representative races of the Western honeybee.

Table 24-2. Differences among Some Important Races of *Apis mellifera*

	A. mellifera ligustica	A. mellifera mellifera	A. mellifera carnica	A. mellifera caucasica	A. mellifera scutellata	A. mellifera intermissa	A. mellifera lamarckii	A. mellifera adansonii
Size	1	> 1	> 1	< 1	< 1	< 1	< 1	< 1
Colour	yellow	black	grey	grey-dark	dark	v. black	striped	dark
Temperament	gentle	defensive	gentle	gentle	defensive	defensive	defensive	defensive
Swarming	low	low	low	low	high	high	high	high
Tongue length	long	long	long	very long	short	very short	very short	short
Propolis	some	some	some	lots	little	little	little	little

CHAPTER VI – 25

HONEYBEE SPECIES & THEIR RELATIVES

Bees are thought to have evolved from a predatory, wasp-like ancestor. But just what is a wasp and what is a bee, in that context? The answer is complex!

The word "wasp" can be applied to a huge diversity of insects. A step into the diversity of insects at higher taxonomic levels explains what is meant (Chapter I – 1). Within the Class Insecta are various Orders. The Order Hymenoptera probably includes at least a quarter of a million species, of which only half or fewer have been described scientifically. Although the Hymenoptera fall into three main categories—the sawflies, the Parasitica, and the Aculeata—the term wasp can be applied to species within each category. The Parasitica and Aculeata have characteristically narrow waists and are placed in the Suborder Apocrita. The Parasitica generally are parasitic on other insects. Their ovipositors are adapted primarily for piercing their hosts and for egg laying. In the Aculeata, the ovipositor usually is adapted primarily for stinging. Looking at the Aculeata in more detail, the boundaries for classification become less clear. The result is that scientific names and what they mean are much debated by insect systematists who research the groups. Already some problems arise: the ants and velvet ants, which are stinging insects, are sometimes included as aculeate Hymenoptera and sometimes not. The three main Superfamilies of the Aculeata in the context of bees are the Vespoidea, the Sphecoidea, and the Apoidea. The Vespoidea and Sphecoidea (wasps) are mostly predatory, and the Apoidea (bees) mostly herbivorous. The Vespoidea include the social yellow jackets and hornets as well as some solitary wasps. The Sphecoidea are almost all solitary, and within this diverse Superfamily are species with characteristics that are shared by bees. Debate has ranged from combining the Sphecoidea and Apoidea into a single, natural (i.e., originated from a common ancestor) taxonomic unit, to initiating a new classification. The big difference between bees and wasps is not so much body form but their food habits. Bees are almost all herbivorous whereas wasps are almost all carnivorous. They can be distinguished by the nature of the hairs on the lower parts of their faces. Sphecoid wasps have shiny hairs, bees do not: their lower facial hairs are either dull or absent. Other useful distinguishing characters for bees are that body hairs are branched and the basitarsus of the hind leg is generally broader than the rest of the tarsal segments (Chapters I – 1 & 3).

The evolution and diversification of bees has been closely linked to that of the angiosperms or flowering plants. The fossil record indicates that flowering plants had become established by the beginning of the Cretaceous period and probably had their origins during the Jurassic period, over 140 million years ago. Geologically one of the oldest fossil bees is *Trigona prisca* (Meliponinae). It is a stingless bee, probably eusocial, and dates from much later. It was found in Cretaceous amber estimated to be 96–74 million years old. From Figure 25-1, it can be appreciated that the origin of the bees must be much older.

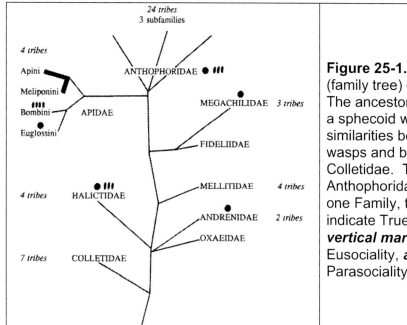

Figure 25-1. Generalized phylogeny (family tree) of the bees (Apoidea). The ancestor is thought to have been a sphecoid wasp. There are great similarities between some such wasps and bees in the Family Colletidae. The Families Apidae and Anthophoridae are now regarded as one Family, the Apidae. **Heavy lines** indicate True Eusociality, *small vertical marks* indicate Primitive Eusociality, **and dots** indicate Parasociality (Chapter I – 1).

Bees are diverse. It has been estimated that there are between 25,000 and 40,000 species. About 16,000 have been described. Bee taxonomy is an active field of research and is receiving more attention because of the importance of bees in pollination and conservation.

The **Colletidae** (plus the closely related Australian family Stenotritidae) are wasp-like in general appearance. They are most abundant as adults in spring or fall. They nest solitarily in the ground, although they may form large nesting aggregations. They do not have special pollen-carrying pouches (corbiculae or scopae), but forage at flowers by mixing nectar and pollen, which they ingest. They carry the forage in the honey stomach.

The **Halictidae**, also known as sweat bees, are diverse. They represent about 60–70% of bee species in Canada. They are mostly ground nesters. Even within the genus *Lasioglossum*, species range from solitary to primitively eusocial. They are important pollinators, and the alkali bee, *Nomia melanderi*, can be managed by provision of special nesting beds for pollination of alfalfa in parts of the western USA (Chapter XI – 49).

The **Andrenidae** (plus the closely related Oxaeidae) are mostly soil nesters. They are important pollinators of some crops and in nature. Many species have short adult lives, flying to and pollinating only one or a few kinds of plants.

The **Megachilidae** is a huge family. The bees are characterized by having their scopae of pollen-collecting hairs on the underside of the abdomen. Most species nest in pre-existing cavities, such as beetle bore holes in wood. Many species make their cells by cutting vegetation (especially leaves and petals), using mud or small stones. They are important pollinators in nature. In agriculture, the alfalfa leafcutter bee (*Megachile rotundata*) (Chapter XI – 46) and orchard bees (*Osmia* spp.) (Chapter X – 48) are raised specially.

The Anthophoridae are considered to be so closely related to the Apidae that they can be placed in a single Family. The Subfamily Anthophorinae is diverse. Many species are robust, resembling bumblebees, but some, such as the dwarf carpenter bees, are small. Some nest in the soil, others in pre-existing holes in stems of wood, and some, such as carpenter bees, actively bore tunnels in wood. They are highly important pollinators in nature, and some are encouraged for pollination of crops (Chapter XI – 49): hoary squash bee, *Peponapis pruinosa*, is one of the most widespread specialist bees, and pumpkin and squash production is indebted to them. *Habropoda laboriosa* can be important as a pollinator of highbush blueberries.

Within the **Apidae** (Figure 25-2), or Apinae, the importance of bees to human affairs through pollination becomes even more pronounced (Chapter X – 44 & Section XI). These bees are sometimes called the corbiculate bees because of the highly specialized hind basitarsus. A corbicula encloses a space for carrying pollen, a scopa consists of pollen-carrying hairs. There are other families of bees that have corbiculae on other parts of the body.

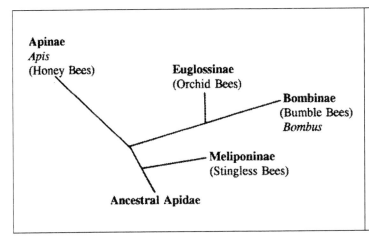

Figure 25-2. A suggested family tree for the Family Apidae (or Subfamily Apinae). The Tribal names (one step lower than Subfamily) would end in "ini" instead of "inae". The position of the branch to stingless bees has been placed above the branch to the bumblebees and orchid bees.

The **orchid bees** (Figure 25-2), Euglossines, have amazing relationships with the flowers of orchids. These bees are found mostly in tropical Central and South America and are fast flying, iridescent green and blue. They often hover, resembling bright jewels somehow suspended over the forest. Their name derives from their exceptionally long tongues that they use to obtain the deeply hidden nectar of the flowers they visit. The males gather orchid perfume on special pads on their fore tarsi. As they hover around the flowers, they pass the oily droplets of scent to the tibia of hind legs. The tibia is highly modified, with a tiny slit to receive the scent and a spongy tissue within to store it. The importance of gathering the scent seems to be in the males' congregating activity, and that is important in attracting females and mating. The orchid scent does not seem to play a direct role in sexual relations. Male euglossines are crucial to the pollination of many species of orchids, but the females do not seem to be involved. Not all euglossines are strictly associated with orchids. They visit the flowers of other plants for sustenance and nest provisions. Some gather floral resins for nest construction.

The **bumblebees** and orchid bees are thought to share common ancestry (Figure 25-2). Tropical forests do not support much diversity: these groups are most diverse in temperate climes. Almost all species are primitively eusocial. Nests are mostly constructed in pre-existing cavities, especially those made by small rodents. There, the founding queen and the workers build a nest of cells in which to store honey, pollen, and raise brood. Bumblebees fall into three categories. The "pouch makers" make a small pocket with each brood cell, and the developing larva feeds from the food the adults place there. The "pollen storers", the brood-attending bees, puncture the brood cells and regurgitate food through the hole. They then seal the hole until the next feeding. The parasitic lifestyle of the cuckoo bumblebees results in species in which the sexual female, the queen, usurps the nest of another species. The founding queen may be killed, or enslaved like the workers she provided. Several species of bumblebees are now used for crop pollination, especially in greenhouses (Chapter XI – 47).

The **stingless bees** (Figure 25-2), meliponines, are eusocial. They build complex nests, mostly in pre-existing cavities such as hollow logs, underground, and in termite nests. Their cells within are made of wax, resin, and sometimes pollen. In general, the brood chamber is arranged in multiple horizontal floors of vertically oriented cells. The queen, or queens (in some species the colonies support several sexually reproductive females) lay their eggs onto the mass of food in each cell. Each larva develops to adulthood in a sealed cell. The meliponines are found throughout the tropical and subtropical world, but are especially diverse in the Americas. Several species are used for honey production in traditional "meliponiculture" from Mexico to Brazil.

Although meliponines are honeybees in the broad sense, the **honeybees** (*Apis* spp.) comprise a natural group (Figure 25-3). Within *Apis*, there are eight well-recognized species. Most are oriental and tropical or subtropical.

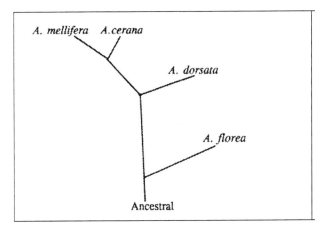

Figure 25-3. Family tree in the genus *Apis*. The sister species of *A. cerana*, *A. koschevnikovi*, and *A. nigrocincta* are all closely related to *A. mellifera*. *Apis dorsata, laboriosa*, and perhaps *binghami* are a similar trio of species. *Apis florea* and *A. andreniformis* are also, sister species.

Species of the *A. dorsata* group and *A. florea* group nest in the open. Their nests are single, suspended combs. Those of *A. dorsata* and *A. laboriosa* (Figure 25-4) are truly immense, often exceeding a metre across where they are attached. A given site, such as a tall tree in the forest, a building, or rock overhang, may support many nests together. Each nest has a brood-rearing area as well as a honey storage area. The latter is located at the top and end of the comb where the foraging bees would first land. Those bees are highly defensive, and may leave the nest en masse to attack intruders. The first sign of alarm is indicated by the workers' making a hissing sound and swaying in waves across the colony, then by the descent of workers so they appear as a cone of bees suspended from the bottom of the comb. It is now time for the bee scientist to leave unless wearing protective clothing! The next stage is massive attack.

Figure 25-4. Left: Numerous colonies of *Apis dorsata* on the branches of a large tree in India. The presence of this bee is regarded as good luck in Asia, so the photograph was used for a greeting card. **Right:** A small nest of *A. florea*, about 10 cm across. The bees have been chased off the comb to expose the honey around the edge and the brood in the centre.

Apis florea and *A. andreniformis* make much smaller nests than do *A. dorsata* and *A. laboriosa*. They are often on shrubs, hidden from view by foliage. The comb wraps around the twig that supports it. The bees have a dance platform atop the comb. These small honeybees can protect themselves by stinging, but the stings are rather weak. To protect their nests from ants, the bees place sticky gum on the twigs just beyond the comb. All those species are exploited for honey by human beings.

The *A. cerana* group and *A. mellifera* (the Western honeybee) are the cavity-nesting honeybees. They nest in pre-existing cavities, which may often be man-made. Within the cavity is an array of more or less parallel combs that comprise the nest (Chapter V – 19). They are all smaller than *A. dorsata* and larger than *A. florea*. *A. cerana* is smaller than *A. mellifera*. Table 25-1 gives the approximate differences in size in the Asian *Apis* species.

Although the diversity of *Apis* is greatest in Asia, not all species have the same geographical ranges. The range of *Apis andreniformis* is more restricted to tropical Southeast Asia, whereas *A. florea*'s range extends westward to include India, Iraq, and part of the Arabian peninsula (Figure 25-5).

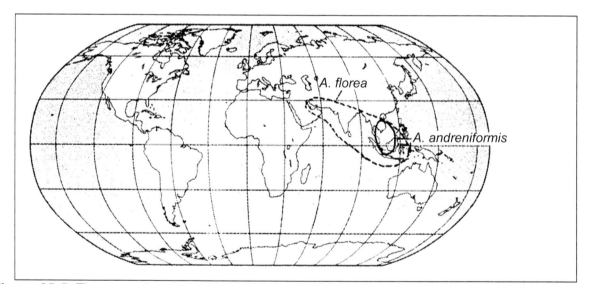

Figure 25-5. The natural range of *Apis florea* (dotted line) and *A. andreniformis* (continuous line).

The range of *Apis dorsata* and its sister species is restricted to tropical and subtropical Asia (Figure 25-6).

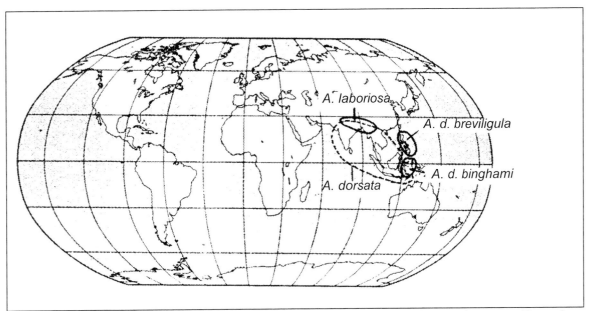

Figure 25-6. Natural ranges of *Apis dorsata* (dotted line) and three other taxa (forms) (continuous lines) showing *A. laboriosa* in north India, Nepal, and Bhutan, and *A. dorsata breviligula* and *A. d. binghami* as distinctive subspecies in Philippines and Sulawesi (Indonesia) respectively.

The cavity-nesting species together share a total range similar to that of *A. dorsata* in the south, but they extend far to the north of the tropics. *Apis cerana* occurs throughout most of China, Korea, and Japan. Figure 25-7 presents some details, but as more is known of these bees, and as the geographic variability in *A. cerana* becomes better understood, it is probable that more subspecies and races will be recognized. In general, the more northerly forms of *A. cerana* are larger than those in the tropics. *Apis c. japonica* is almost as large as *A. mellifera*. Colour forms are known, sometimes called dark (or mountain) bees versus pale bees. Those may be seasonal variations.

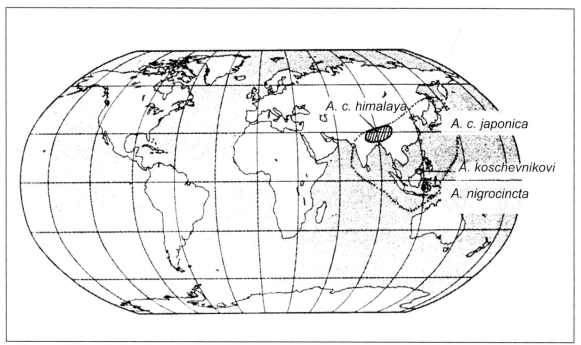

Figure 25-7. Natural ranges of *A. koschevnikovi* (in Southern Malaysia and Western Indonesia) and *A. nigrocincta* (on the Island of Sulawesi, Indonesia) are totally within the range of *A. cerana*. Races of *A. cerana* are recognized from the Himalayan region (*A. c. himalaya*), Japan (*A. c. japonica*), and perhaps India and Sri Lanka (*A. c. indica*). The relationships between these races are a fertile area for research.

With such diversity of similar bees in parts of tropical Asia, the question arose as to how they prevent cross breeding. Work in Sulawesi with *A. cerana* and *A. koschevnikovi* showed that the mating times of the two species were separate (Figure 25-8). Similar research has shown differences in mating times between *A. cerana* and *A. nigrocincta* in Sulawesi, Indonesia. The behavioural differences are reinforced by differences in the structure of the males' intromittent organs. Those anatomical differences are presumed to work in a manner similar to a key in a lock, where the lock is the vaginal structure of the females. Together, the behavioural and anatomical differences constitute **reproductive isolating mechanisms** that are the basis for the concept of species, i.e., individuals from different species are unable to mate with each other to produce fertile offspring.

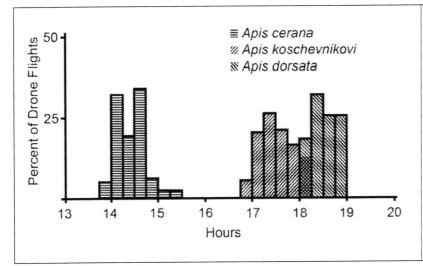

Figure 25-8. Drone flight times of *Apis cerana* (at about 2:30 p.m.) and *A. koschevnikovi* (at about 5:30 p.m.) and *A. dorsata* (at dusk at about 6:30 p.m.) in Sabah, Malaysia.

	Apis cerana	*Apis dorsata*	*Apis florea*
Worker body size (partially loaded)	54 mg	155 mg	32 mg
Nest site height visibility dispersion	Cavity Low (<2–3 metres) Conspicuous Spaced	Tree limb or cliff High (mostly >15 metres) Conspicuous Often clustered	Branch of shrub Low (mostly <5 metres) Hidden Spaced
Colony population aggressiveness movements foraging area foraging range mass of entire colony	7,000 Low Stable Small (<10? km^2) About 1 km 0.4 – 1 kg	37,000 High Migratory Large (>300? km^2) About 5–10 km 6.0 kg	6,000 Low Local Small (<3? km^2) About 0.5 km 0.2 kg

Table 25-1. Comparison of three honeybee species of Asia. *A. mellifera* (not shown) is intermediate in size between *A. dorsata* and *A. cerana*.

Section VII

Senses: Bees' Perceptions of the World
Vision & Light, Smell & Taste, Hearing, Touch & Position, Temperature, Humidity, Time, Magnetism & Electricity

The honeybee's brain, with its 850,000 cells, receives information from a marvelous sensor array. Some aspects of that array have been described in brief in Chapter I – 2, but in this section, their workings are described.

The five **eyes**, of two kinds, have different **photoreceptive** tasks. The **ocelli** are little more than light meters. The **compound eyes** need bright light to gather information on images: **colours, colour patterns, sizes, shapes, symmetries, edges, and movements**. The compound eye works in a way completely unlike our own camera eye. Each of the tiny facets of a compound eye can receive information in three wavelengths (or colours; note that light rays are not coloured): **ultraviolet, blue,** and **green**. The mechanisms of seeing colours, and remembering colours of objects, are complex. This chapter presents a general account of what is known, but it is much simplified, even with the detail included. Try to understand the concepts, but the ideas are difficult and new to most biologists. Nevertheless, to gain an understanding of how insects, and particularly bees, see objects of interest to them, such as flowers, human beings must try to record images in accord with bees' vision. The results are startling. The compound eyes also sense the **polarization of daylight**. That is a major advantage to bees navigating on cloudy days.

Smell and **taste** are chemical senses. Most tasting is done on the mouthparts, but the antennae can taste as well as smell. Those two compound elements of the sensor array are, themselves, covered with microscopic sensors called **sensillae**. There are several generic types that respond to types of stimuli: **chemicals, touch, heat, humidity, carbon dioxide,** and **vibration**. The senses of touch and vibration are mediated by **mechanoreceptors**. Those are also involved in providing to the central nervous system information about the relative positions of body parts, **proprioception**. The organs of **hearing, Johnston's organs,** are located on the antennae. Bees were thought deaf until recently!

Bees can tell and measure **time**. That is an important skill when navigating by the sun as it traverses the sky over hours of foraging. They also remember at what time events take place from day to day.

Bees seem to be sensitive to **magnetism** and **electricity**, but how that works is not understood.

CHAPTER VII – 26

VISION & LIGHT

The capacity of organisms to sense light extends to plants. However, their growth responses to light hardly qualify as vision. Vision as a sense can be restricted in its meaning to animals and their use of special organs—eyes—to sense light. Light may be sensed in different ways:
- Presence (above a certain threshold) or absence (below that threshold);
- Discrimination of wavelengths, ultimately represented by colour vision;
- Discrimination of images (i.e., shapes and sizes) through sharpness of vision, i.e., the resolving power of the eye;
- Polarization, i.e., the angle at which light rays differentially pass through a medium (usually air or water);
- Speed of reaction of the optical sensors that allow for sensing of movement.

The eye is a highly specialized organ of **photoreception**, a process that involves the conversion of quanta of light energy into nerve action potential. There are three major types of eye:
- Simple eyes are generally cups of photosensitive cells that react to light; they can detect brightness and direction (Figure 26-1);
- Camera eye of vertebrates;
- Compound eye of arthropods, including insects.

The three **simple eyes** or **ocelli** of the honeybee are located on the top of the head (Chapter I – 2). Ocelli **can monitor light intensity** (Chapter V – 17), **period of exposure to light**, and **wavelength.** Each has one lens for the entire retina. That lens refracts incoming light to about 800 photoreceptor cells, but the retinal stimulation cannot form an image.

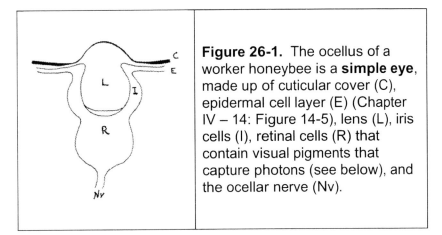

Figure 26-1. The ocellus of a worker honeybee is a **simple eye**, made up of cuticular cover (C), epidermal cell layer (E) (Chapter IV – 14: Figure 14-5), lens (L), iris cells (I), retinal cells (R) that contain visual pigments that capture photons (see below), and the ocellar nerve (Nv).

For interest's sake, and to understand the differences and similarities between the two kinds of **image-forming eyes**—the **camera eye** (of vertebrates and cephalopods (octopus and squid) and the **compound eye** of arthropods—it is worth a slight excursion into how our own eyes work (Figure 26-2).

In the **camera eye** (e.g., of human beings), light enters the eye through the cornea and aqueous humour, is blocked by the iris (which adjusts according to the brightness of the light) and enters the pupil. The light then goes through the **lens** and the vitreous humour. The human eye changes focus for near and far vision by changing the shape of the lens. The image is cast, upside down, on the **retina.** Our brains adjust the image so that it is appreciated in its true orientation.

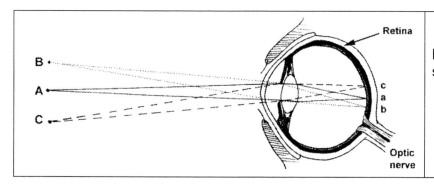

Figure 26-2. Diagrammatic section of the human eye.

In the retina are the receptor cells, which are modified dendrites (Chapter III – 9) of two types of nerve cells: **rod cells** and **cone cells.** The **rods** (125 million) are not colour sensitive (i.e., they are **achromatic**). In the camera eye, the rods are integrated into a system that is receptive to light at different intensities. The image perceived is in a form analogous to a black and white (gray tone) photographic image. The **cones** are much fewer (6.5 million) and are colour sensitive. They are especially numerous at the fovea (the critical area of the retina for colour vision and image perception). The cones are of three types, receptive to the colours blue, green and red (with lots of overlap), and constitute a system by which coloured images may be perceived **trichromatically**. In the retina are integrating neurones that start the process of visual processing of optically gathered information.

Both rods and cones contain **visual pigments**: rhodopsin and iodopsin. These pigments comprise retinal (a vitamin A aldehyde) and proteins (called opsins). Rhodopsin, or visual purple, is the molecule that captures quanta of light (photons) in achromatic vision. Iodopsin is the visual pigment for colour vision.

After a quantum of light enters the eye, enters a retinal cell and interacts with, for example, rhodopsin, the pigment changes into lumirhodopsin. That molecule is unstable and breaks down rapidly. In doing so, its energy is released and excites the nerve cell.

The nervous impulse travels via the dendrite (where the transducer system is located), down the axon to the brain. In the brain, the information for all the retinal cells is synthesized and results in our visual appreciation of the world around us.

The breakdown products are resynthesized to rhodopsin, and that takes time. A light flash of a millionth of a second lasts one tenth of a second as an optical image in the eye. That persistence of the image allows perceptual fusion of flickering images, such as in movies, TV, and fluorescent lights. Thus we see still images flashed rapidly on a movie or TV screen as motion. The flicker fusion frequency for human beings is about 40 images per second.

Colour vision is less well understood.

Colour can be defined as visual perception associated with the various wavelengths in the visible portion of the electromagnetic spectrum of light. The perception of colour is a complex neurophysiological process. As Isaac Newton remarked, the light rays are not coloured, but they are perceived that way because they cause the sensation of colours.

The colour of light of a single wavelength or of a small band of wavelengths is known as a pure spectral colour or hue, but most colours in nature are not as pure. Within the three kinds of cones, there is a lot of overlap in the wavelengths that excite them, e.g., green cones are excited by light from 450 to 675 nanometers (nm): light at 450 nm would appear blue, and at 675 nm would appear red. Nevertheless, each cone type has its own peak sensitivity: blue cones at 436 nm; green cones at 546 nm; and red cones at 700 nm.

Red, green, and blue are the three **primary colours**, thus this type of vision is known as **trichromatic colour vision**. If one looks closely at printed colours in a newspaper, one can notice dots of these three colours. Colour mixing on TV screens works similarly.

Not all animals process coloured images using exactly the same three colours. Some, such as some birds and fish, use four primary colours: ultraviolet is the additional primary. Some animals have only two, as for example, in people who are red–green colour blind. In **achromatic** vision there is no colour, as for nocturnal animals.

Honeybees, and all insects (except a few that have no eyes), have two **compound eyes**, located on the sides of the head. Each of these has many **ommatidia** or **facets**, each with its own lens. In honeybees, each compound eye has about 4–5,000 facets in the worker, 7 - 8,000 in the drone, and about 3,500 in the queen. They appear as a curved mosaic of hexagonal tiles (Figures 26-3 & 26-4).

Figure 26-3. The compound eye of a worker honeybee. The individual facets can be seen, along with long mechanosensory hairs that stand between the facets, inserting at the corners of hexagonal facets.

Each facet looks out at an angle of about 1° where they are the most compact (an area called the fovea), and 2–4° where they are not as compact. Single facets, accepting information from just that small region of perception, gather the image. The image is gathered and processed in a way that can be appreciated by considering **mosaic** pictures.

The size of the tiles making up a mosaic constrains the amount of detail. A small picture made of a few tiles is difficult to appreciate in comparison with a large picture made of many small tiles. Then, the distance one is from a mosaic changes one's ability to discern the picture. If one is close to a large picture of many relatively small tiles, the total picture cannot be seen. Those same problems constrain the ability of insects to see objects. The ability to see detail is **resolving power**. For human beings that is a cone of about 1 second solid angle, but for the honeybee worker it is about 1° in the fovea. That indicates that an object of about 10 cm in diameter should be visible to a single ommatidium from about 5.5 metres distance. Experiments have shown that perception is more complex. Depending on the colour (see below), an object of that size is not recognized until the honeybee is about 1 metre or less from it. The mechanisms of colour vision, and of perception of the contrast of objects against their backgrounds in the green part of the spectrum, are important.

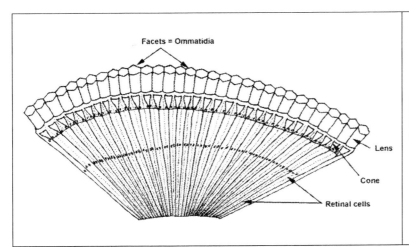

Figure 26-4. Diagrammatic structure of part of the faceted eye of a honeybee.

The compound eyes of honeybees are organs for colour vision, shape and pattern vision, contrast vision, and detection of motion. All those modalities are interrelated in how the compound eye works.

Light enters each facet through the corneal lens and passes into the cone lens (**crystalline cone**). The crystalline cone is surrounded by **primary iris cells** that reflect incoming light to the **retinula cells**. Those cells contain visual pigments that work as described above for vertebrates, and so stimulate nervous impulses to the brain.

Each **ommatidium** contains **nine retinula cells** sensitive to different colours. The cells are twisted about a structure called the **rhabdom** that runs the long axis of the ommatidium. The **rhabdom** is formed by fine membranous interdigitations of the individual visual cells. The short cell (number 9) is present only near the nerve fibres that conduct impulses. Three types of visual cells, **green**-sensitive, **blue**-sensitive, and **ultraviolet**-sensitive, are present. Thus, the colour vision system is **trichromatic**. The retinula cells are surrounded by secondary iris cells that prevent the scatter of light from one ommatidium to the next. In some insects, the secondary retinula cells can adjust according to light intensity (Figure 26-5).

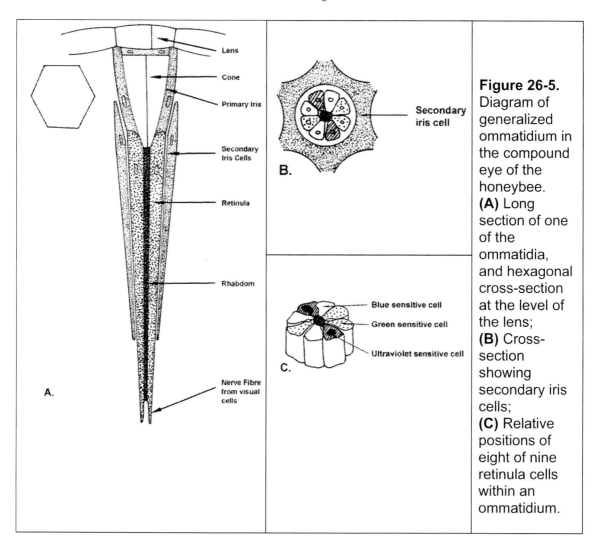

Figure 26-5. Diagram of generalized ommatidium in the compound eye of the honeybee. **(A)** Long section of one of the ommatidia, and hexagonal cross-section at the level of the lens; **(B)** Cross-section showing secondary iris cells; **(C)** Relative positions of eight of nine retinula cells within an ommatidium.

Trichromatic colour vision in bees does not use the same three wavelengths as in human beings. Bee vision includes ultraviolet, blue, and green, but human beings cannot see ultraviolet, and bees cannot see red. That does not mean that what human beings see as a red object is invisible to bees, just that it is black (unless it reflects ultraviolet as well). The visual spectrum of honeybees is shifted to shorter wavelengths in comparison with that for people (Figure 26-6).

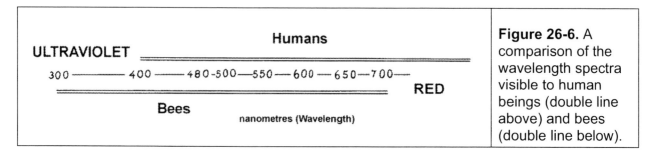

Figure 26-6. A comparison of the wavelength spectra visible to human beings (double line above) and bees (double line below).

In trichromatic colour vision, mixing lights of different primary wavelengths create **secondary colours.** The system can be represented by pairs of complementary colours that lie opposite to each other in a "colour wheel" (Figure 26-7). Mixtures of appropriate amounts of three complementary colours appear indistinguishable from white light.

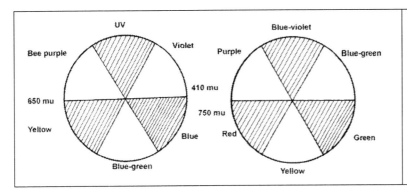

Figure 26-7. Colour circles that represent primary (shaded) and secondary colours for insects (**left**) and human beings (**right**).

The colour circles are a simplistic way of presenting colours. To determine how colours, and differences between colours, are perceived by honeybees, an elaborate series of multiple-choice experiments were made. The result was a new understanding of colour vision in insects through **colour opponency coding** put forward by Werner Backhaus and co-workers at the Institute for Neurobiology at the Free University of Berlin. Simply put, worker honeybees were trained to go to one of 12 colours presented outdoors on a vertical wheel, as shown diagrammatically in Figure 26-8. Once the bees had learned to associate a sugar syrup reward with a particular "training" colour, the training colours were removed and the bees were exposed to 12 other colours. The researchers made careful observations of each bee's choices and the data were analyzed.

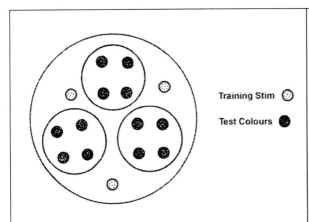

Training Stim ◎

Test Colours ●

Figure 26-8. Simplified sketch of the training site for multiple-choice experiments showing the "training colour" and the "choice colours". The whole apparatus could be rotated, and the discs of choice colours also rotated separately. That avoids the potential problem of the bees' recognizing position as the cue for rewarding syrup.

The choices made by the bees were presumed to be related to the similarity of the training colour to the colour chosen (i.e., the more similar the choice colour to the training stimulus, the more visits it would receive). By testing thousands of individual workers against many combinations of training colours and choices, the dimensions of the colour space could be calculated. Moreover, the way in which bees' colour perception works could be explained.

The simplest explanation results in a colour opponency colour space that can be represented by an oval, rather than a circle. A hexagonal colour space can also be used, and again the similarity to the colour wheel is evident. The shapes of the colour spaces reflect how the three kinds of receptors operate in each ommatidium of the compound eye (Figure 26-9).

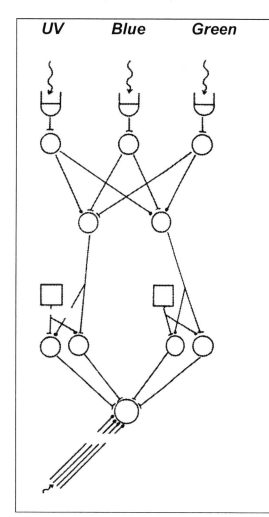

UV   Blue   Green

Figure 26-9. A simplified explanation of colour opponency coding and colour choice by honeybees. Light (as waves) enters ommaditidia and stimulates retinula cells (U-shaped) for any of UV, blue and green. The nervous impulse is generated to the monopolar cells (first set of circles). Those, in turn, send nervous impulses to the next layer of cells, the colour opponency coding (COC) cells. Those impulses are either excitatory (from blue) or inhibitory depending (from UV), or both (from green) on the COC cell they reach. The net effect of the spectral properties of the optical stimulation is conducted to the brain. There, it is presumed that information stored in memory is compared with that being processed when the bee makes its choice. The results of those comparisons initiate the bee's behaviour. The explanation becomes more complex when differences and similarities in background colours, choice colours, and colours stored in memory are considered.

In short, information on the colour of an object of interest to an individual bee is first processed by the three kinds of retinula cells, then coded on the two colour opponency neurones, all the while with reference to background colours coded in the same manner by adjacent ommatidia. That information must then be compared with information stored in the bee's memory, referred to as **triadic comparison**. Higher brain function leads to the bee's decision (i.e., choice as to what colour is most similar to that stored in memory as associated with syrup reward).

The evidence for the system resides in actual electrophysiological recording from retinula cells, monopolar cells, and colour opponency coding cells. Deeper than that, single cell electrophysiology has yet to penetrate. Nevertheless, bee behaviour, and some aspects of the brain's anatomy, support the model in many respects.

The results of electrophysiological experiments, and those that lead to the COC model, also allow for an understanding of spectral discrimination. Figure 26-10 presents the relative sensitivity of each of the receptors to the primary wavelengths of light. Each receptor has its maximum sensitivity set at 1. It is known that the UV receptor is more sensitive to light than is the blue receptor, and the green receptor is the least sensitive. That pattern follows the amount of light in those wavelengths present in normal daylight. Just as human eyes can adapt to ambient light of different colouration (e.g., at dusk or dawn when daylight is richer in red light, or for indoor lighting), so too can the honeybee's eyes.

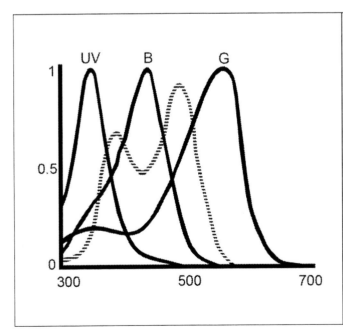

Figure 26-10. Spectral sensitivity curves for the three primary colour receptors (ultraviolet, blue and green) of the honeybee's ommatidium (solid lines) and the spectral **discrimination function** (dotted line). Note that the eye is more sensitive to subtle changes in colour when at least two receptors are involved in coding. It is interesting that it is at those peaks that one finds most flower colours (Figure 26-11).

To understand the colours of flowers as insects, particularly bees, may see them requires recording ultraviolet, blue and green reflections, and disregarding red reflections. There are various ways to do that, including by photography. By measuring the amount of light reflected from objects of interest to bees, such as flowers and their backgrounds, and placing that information into the various, but similar, colour spaces that have been used (mostly triangles and hexagons), three important points emerge: (1) that there are many more discrete floral colours in the bee's colour vision system than in our own, (2) that the discrete floral colours occupy about double the area of bee's colour space that they do in the human colour space, and (3) that most flowers occupy parts of the colour space where colour discrimination is the greatest (i.e., in parts where at least two different retinula cells are stimulated) (Figure 26-10).

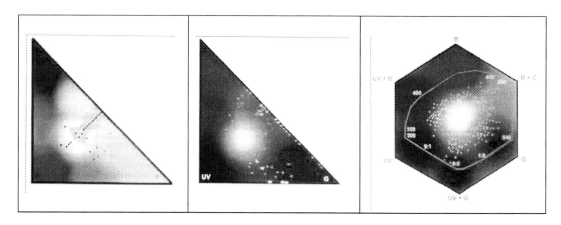

Figure 26-11. The **left** and **centre** triangles (simple colour spaces) compare the colours of flowers of common weeds in the human visible spectrum and the insect visible spectrum. Flowers that are white to human beings reflect all three primary colours, so fall in the centre of the colour triangle. Flowers that are yellow to human beings reflect equally red and green, so they fall halfway along the red-green axis. Bluish and purplish flowers tend to be pale, so fall toward the blue primary and toward the blue-red axis. The bee-colour vision triangle (**centre**) ignores red, but includes UV. The colours of the flowers spread more over this colour space than the human colour space. There are almost no UV flowers, although several reflect both UV and green (i.e., bee purple of Figure 26-7) and are close to that axis. Many more flowers are represented in the more complex and realistic hexagonal colour space of Lars Chittka (**left**), but the same pattern of many discrete flower colours is presented. The large numbers of flowers with colours that reflect in two bee primary visual wavelengths (e.g., especially B + G) would be especially discriminated (see Figure 26-10). In the "bee-white" centre of both the simple triangle and the hexagon, there are almost no flowers, but that is where the colours of backgrounds, such as leaves and soil, fall. Thus, it seems that floral colours and bee colour vision accord with each other.

Human beings have a far greater ability to see details of flowers' shapes, but bees have a far greater ability to detect differences in the colours of flowers. Figure 26-12 illustrates a prickly pear cactus flower as it appears to human beings and as it may appear in the insect's eye.

Figure 26-12. The flower of a prickly pear cactus is uniformly yellow to the human eye, but is intricate in its form. The flower is also uniformly coloured to the insect eye, but when rendered as hexagonal "pixels" to represent mosaic vision, the intricate detail is lost. It is estimated that the pixelated resolution corresponds to the bee's being about 7 to 8 cm from the flower.

Although bees' eyes have much less resolving power than do human eyes, they are able to **orient to sizes and shapes**. It appears that such tasks, and visual sensitivity to **motion**, **contrast**, and **distance** (see Chapter V – 17) are mediated by the green receptors of the ommatidia. The mechanisms are not well understood.

The results of some early experiments showed that honeybees easily learn to distinguish each of the upper figures from any of the lower ones but cannot distinguish between the shapes *within* the upper or the lower group (Figure 26-13).

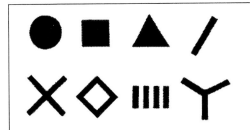

Fig. 26-13. Shapes used in a classical experiment of shape discrimination by honeybees. These shapes are used today on hive entrances and leafcutter bee domiciles with a view to aiding the bees' orientations.

Later, though, E. E. Leppick noted that honeybees could discriminate between shapes representing floral forms, notably a Bar, a Vee, and radiating patterns of 3 (a Y), 4 (a cross), 5, 6, 8, and "many" arms. He referred to those as "figure numerals". Since then, a number of researchers have tested the abilities of honeybees to discriminate between patterns and shapes, and they have been found to have remarkable abilities. Figure 26-14 presents a sampling of those tested.

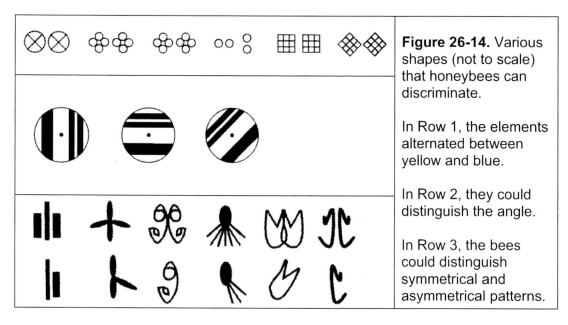

Figure 26-14. Various shapes (not to scale) that honeybees can discriminate.

In Row 1, the elements alternated between yellow and blue.

In Row 2, they could distinguish the angle.

In Row 3, the bees could distinguish symmetrical and asymmetrical patterns.

Given that honeybees seem to have such limited visual resolution as dictated by the mosaic vision of the compound eye, yet can discriminate such subtle differences in shape, it is fair to ask "how do they do it?" They scan objects of interest as they approach closely. Figure 26-15 illustrates scanning behaviour by honeybees flying as they approach some patterns with which they have been presented.

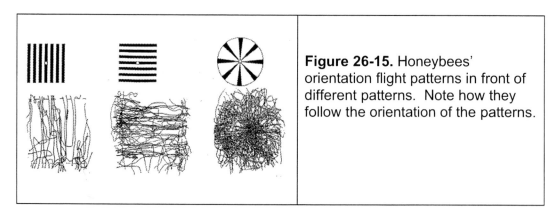

Figure 26-15. Honeybees' orientation flight patterns in front of different patterns. Note how they follow the orientation of the patterns.

It is often stated that honeybees are attracted to divided shapes more than to uniform ones. That is not true. Experimental results show that complete discs are more visible from greater distances to honeybees than are dissected shapes of the same diameters. The apparent greater attractiveness may reside in the close-in orientation cues that a divided shape may present that a uniform disc lacks.

All in all, honeybees' capacities to distinguish patterns, and to orient so precisely to them, is presumed to be mediated through **flicker fusion**. The structure of the compound eye, with its facets each looking out to a different part of the environment, is admirably suited to noticing differences between the facets. Honeybees' eyes have a flicker fusion frequency of 1/150 to 1/200 second (experimental results show that they can register separated stripes moving past them at 150 to 200 stripes per second!). Thus, they have a much higher appreciation of motion than do human beings. That motion may be of objects moving around them, or, in the case of the scanning behaviour, their movement in relation to stationary objects. Human beings' flicker fusion frequency is about 1/30 second, so only at frequencies above that do objects appear to move smoothly rather than in flicks (as in the old movies, "the flicks"). Apart from honeybees' using visualized relative motion in orientation, they use the same mechanism for *monitoring speed* over the ground (see Chapter V – 17 and the optical flow theory in distance estimation) and in depth perception. In the latter, they would be able to appreciate the increasing number of ommatidia stimulated as they approached an object, and be able to combine information from each eye through stereoscopic vision.

The correlates of colour vision, motion detection, and pattern recognition clearly are important to honeybees' foraging at flowers. The general colouration of flowers is different from that of their backgrounds, and as has been noted, flowers in a given environment present a greater array of colours in the bees' visual system. The detailed colour patterns of flowers also have roles in the orientation of bees to nectar and pollen.

Nectar Guides may be patterns of colour, scent (Chapter VII – 27), or structure (Chapter VII – 29). Coloured Nectar Guides may or may not be visible to human beings. Many yellow daisy-like blossoms (as in the Aster family) have ultraviolet reflections which human beings cannot see (Figure 26-16).

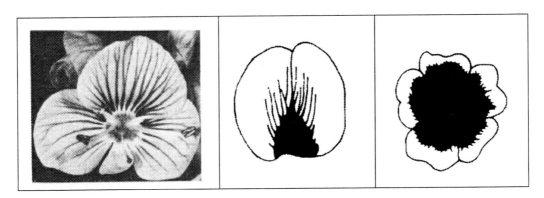

Figure 26-16. Some Nectar Guides: on *Veronica* (**left**), a common lawn weed that has blue flowers with dark blue guides; some legumes (**centre**) with yellow flowers have parts that strongly reflect ultraviolet light, as do flowers of some ornamental cinquefoils (**right**).

The function of Nectar Guides has been shown experimentally. Karl Daumer reversed the florets of a yellow daisy blossom and found that honeybees oriented themselves to the ultraviolet-reflecting outer edge, then moved to the centre on the normal bloom, but reversed their movements when the florets were reversed (Figure 26-17).

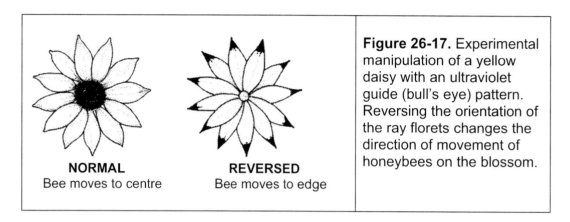

NORMAL
Bee moves to centre

REVERSED
Bee moves to edge

Figure 26-17. Experimental manipulation of a yellow daisy with an ultraviolet guide (bull's eye) pattern. Reversing the orientation of the ray florets changes the direction of movement of honeybees on the blossom.

The position of the sun in the sky is crucial to honeybees' navigation (Chapters V – 17 & 18), yet they are able to navigate on cloudy days when the sun is invisible to human beings. Bees, and other insects such as ants, can detect the **polarization of daylight** and thereby know where the sun is in the sky. Light becomes polarized as it travels through the Earth's atmosphere from the sun. In the immediate direction of the sun, the light is the least polarized: the light waves oscillate in all directions perpendicular to the direction they travel (Figure 26-18). As sunlight enters the atmosphere it becomes scattered, and in the process the oscillation angles become constrained. Thus, at the arc in the sky 90° away from the position of the sun, the light is the most polarized. The result is that the symmetrical pattern of polarized light around the sun is an accurate **sky compass** for orientation.

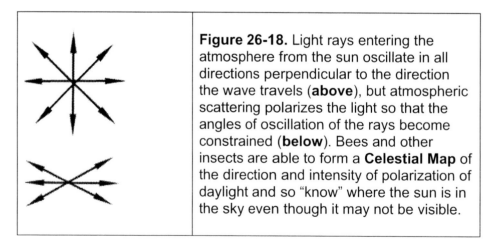

Figure 26-18. Light rays entering the atmosphere from the sun oscillate in all directions perpendicular to the direction the wave travels (**above**), but atmospheric scattering polarizes the light so that the angles of oscillation of the rays become constrained (**below**). Bees and other insects are able to form a **Celestial Map** of the direction and intensity of polarization of daylight and so "know" where the sun is in the sky even though it may not be visible.

The compound eye has a special area with a narrow band of facets along the dorsal rim (the **POL area**). Those ommatidia are specialized in remarkable ways to be able to sort out the angles of polarization of the light they are receiving. Features such as the tiny interdigitations (microvilli) of the retinula cells, fine canals in the corneal lenses, ommaditial sizes and orientations, sensitivity to ultraviolet light (more polarized than other wavebands), and information from each eye combine together with neuronal integration and information processing in the brain to allow bees a remarkable sensory capability that is denied to human beings.

CHAPTER VII – 27

SENSES OF SMELL & TASTE

Smell (**olfaction**) and taste (**gustation**) are chemical senses. They are difficult to separate, even in human beings. Our appreciation of food is through smell and taste together and is mediated by **chemoreceptors** of **olfactory** and **gustatory** organs. Different animals have greater or lesser sensitivities to smells and tastes of different chemicals. Bloodhounds have exceptionally powerful olfactory capabilities, but most birds almost none. The arrays of smells and tastes cause different reactions in different animals so that a smell or taste that is relished by one animal may be repellent to another. Worker honeybees are up to 100 times more sensitive to flower, wax and other bee-significant odours than are humans. Chemosense is crucial to colony coordination, (pheromones; Chapter V – 16), foraging (Chapter V – 18) and other behaviours.

Unlike vertebrates, insects are enclosed in an exoskeleton that would prevent penetration of sensory information if special areas on it did not exist (e.g., Chapter VII – 26 for reception of light). Chemical information must get inside to stimulate chemosensory neurones. On the exoskeleton are specialized microscopic structures made of thin cuticle with neurones beneath for sensory transduction. Those structures are called **sensilla** (**sensillum** is the singular). They occur on many parts of the body, but especially on the antennae, mouthparts, and feet. They come in various forms (Figures 27-1 & 27-2), and one type of sensillum may transduce more than one kind of stimulus, i.e., not just chemical or touch, but both. On a given area of the body, the sensilla of a certain type may be specific to a particular stimulus, such as bending (Chapter VII – 29).

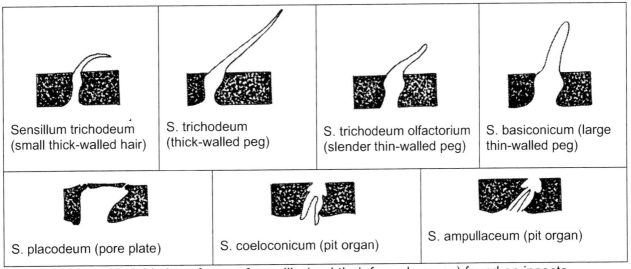

Sensillum trichodeum (small thick-walled hair)	S. trichodeum (thick-walled peg)
S. trichodeum olfactorium (slender thin-walled peg)	S. basiconicum (large thin-walled peg)
S. placodeum (pore plate)	S. coeloconicum (pit organ)
	S. ampullaceum (pit organ)

Figure 27-1. Various forms of sensilla (and their formal names) found on insects, including the honeybee.

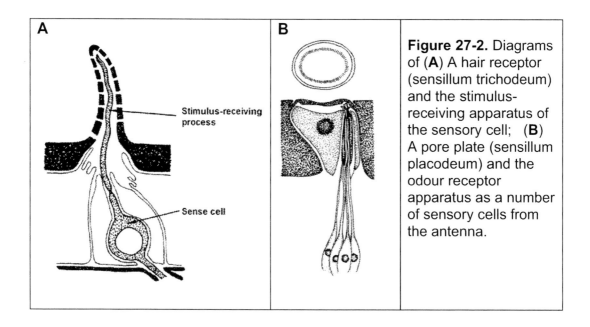

Figure 27-2. Diagrams of (**A**) A hair receptor (sensillum trichodeum) and the stimulus-receiving apparatus of the sensory cell; (**B**) A pore plate (sensillum placodeum) and the odour receptor apparatus as a number of sensory cells from the antenna.

The sensilla with pores (i.e., trichoid, basiconic, placoid) are believed to be associated with smell and taste. Molecules on the air, or on surfaces, penetrate the pores and interact with the nerve's dendrites within the sensillum. It is thought that the molecular structure and physical and electrical properties interact with the membrane of the dendrite in a kind of physical and electrical key and lock manner so that molecules with certain properties cause activation of the dendrite and generation of a nervous impulse that eventually reaches the central nervous system (Chapter III – 9).

The worker honeybee has about 40,000 sensilla on each antenna. There are estimated to be at least 3,000 hairs and pegs with pores (sensilla trichodea) and between 2,700 and 6,400 pore plates associated with chemosensory transduction. There are also a large number of hairs and pegs that lack pores. On the queen honeybee's antenna, the number of pore plates is about half that on the worker, but on the drones' larger antenna there are about 15,000–19,000. Remarkably, each plate is about 12 microns across, has 18 or so cells beneath it, has 3,500 pores, and each pore is 15×10^{-6} mm across.

The complex sensilla coeloconicum or ampullaceum (Figure 27-1), with pits and simple or fluted pegs within, is thought to be sensory for temperature, humidity, and carbon dioxide.

Sense of Smell (Olfaction)

Behavioural experiments have been used to examine the sense of smell in honeybees. In such experiments, honeybees are challenged to distinguish a training scent, associated with a syrup reward, from another scent presented at the same time and in the same way. Although it seems that, except for special cases, honeybees' ability to perceive scents is a little better than our own, they have a similar ability to discriminate between scents. It has been estimated that human beings can differentiate about 2,000 scents. In trials of 1,816 odour pairs presented to honey bees, they could discriminate between the two 1,729 times.

Honeybees orient to odour flow. Figure 27-3 shows that a honeybee with one antenna removed orients to odour in a Y-maze olfactometer by turning in the direction of the remaining antenna. In more elaborate experiments in which, after training, the antennae were left intact but were anchored so that they were crossed (the left antenna on the right side, and the right antenna on the left), the bees turned away from the training scent at the point at which they had to choose.

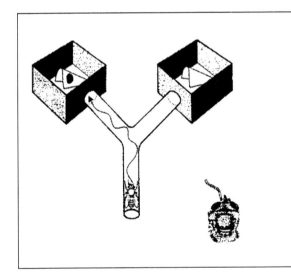

Figure 27-3. Representation of the experiment which shows how a bee without a right antenna and with a left antenna follows a typical oscillating path to scented paper. If the scent were coming from the right-hand side of the maze, at the point of choice at the divergence of the tubes, the bee would become confused. It would follow, based on information from the left antenna, the left (unscented) tube for a short way. Then it would probably back up before taking the other tube.

Trials have also shown that honeybees can remember sequences of scents. In a four-armed olfactory choice chamber, honeybees were trained to associate the following scents in the following order: rosemary-thyme-fennel, then were given choices of the same scents, but in different and the same orders. Figure 27-4 presents the results of those experiments. Most of the bees oriented to the training sequence, some oriented to the first scent (rosemary) but few continued, and visits to the ends of the arms with other scents at the start were few.

The fact that bees can orient correctly to sequences of odours has implications for their abilities to navigate in complex landscapes of visual and olfactory information

(Chapters V – 17 & 18). Nectar Guides (Chapter VII – 26) come as colour (Chapter VII – 26), texture and structure (Chapter VII – 29), and scent patterns. The "mini-landscape" of a flower contains information for the main sensory modalities. There are known differences between the chemicals of the petals versus the stamens of roses. Bees orient to those patterns. Moreover, on petals of some flowers there are lines of scents to complement the colour and texture patterns as guides.

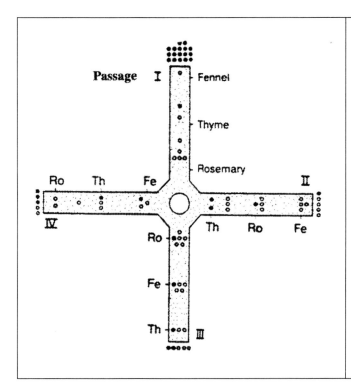

Figure 27-4. The experiment in an odour arena demonstrating the learning ability of honeybees for scent. Bees enter through the opening in the middle. Then most of them proceed through Passage I, containing the combination "rosemary-thyme-fennel" to which they were trained. The open circle indicates one event, the black circle five events.

Sense of Taste (Gustation)

It is generally believed that there are four main components of taste for human beings and for honeybees: sweet, salty, sour, and bitter. Even so, many different and subtle tastes can be appreciated. Taste can be distinguished from smell in that the gustatory sensors are in contact with the chemical stimulus, but the distinction is rather arbitrary when one realizes that airborne chemicals contact olfactory receptors, too. Thus, the matter of whether or not the chemical sensed is vapourized makes the distinction. Some chemicals, such as sugars, do not vapourize and so can be only tasted. In bees, sensory taste structures are located on the **mouthparts, antennae, and feet.** The sensilla used in taste detection are hairs and plates with pores, and the same mechanism applies as for smell.

For honeybees, the minimum amount of sugar in a solution needed for detection of what human beings would call "sweetness" is 3% (2–4%). In human beings the threshold is much lower, at about 0.3% sugar. At first the difference seems the reverse of what would be expected, but then the energetics of foraging offers an explanation (Chapter V – 18). A foraging honeybee that expended energy on collecting dilute nectar would likely inefficiently expend more energy than it gained by carrying mostly water to the hive.

Taste is the sense used for detection of sugars, salts, and other chemicals, such as some pheromones that are not volatile (Chapter V – 16). It is no doubt used for detection of chemicals that do vapourize and can be smelled as well.

Other Chemical Sensitivity

Honeybees are highly sensitive to **carbon dioxide** (CO_2) and can react to 1% differences in its concentration. That is important in maintaining fresh air in the close confines of the hive and clustered colony (Chapter V – 19).   Also related to maintaining a "healthy" atmosphere in the hive is humidity (Chapter V – 17) and temperature (Chapter V – 19). Honeybees are highly sensitive to **humidity** and to **temperature** (Chapter VII – 30). They can detect changes in relative humidity of about 5%, and changes in temperature of less than 1°C. The sensors involved are mostly on the antennae and are made up of sensilla ampullaceae and sensilla coeloconica (Figure 27-1), of which there are about 300 on each antenna.   Electrophysiological recordings from the neurones of some of these sensory arrays have shown three kinds: a moisture receptor, a dryness receptor, and a temperature receptor grouped together.

CHAPTER VII – 28

HEARING

Hearing, or the perception of sound, is a nervous response to oscillation of pressure in the medium in which the animal lives, whether air or water. Sound is generated by the molecules of the medium being compressed and rarefied. A speaker on a stereo system vibrates, causing those changes in the air in front of it. Those waves, like waves on a pond disturbed by a rising fish, are propagated through the air. Hearing is the perception of those waves of compressed and rarefied air.

In human beings, the main sensory element of the ear is the *vestibulo-cochlear* apparatus, but before sound waves are converted to perceptions they must be processed. First, the **external ear** funnels sound onto the eardrum (tympanic membrane). The drum vibrates, converting the waves to mechanical vibrations that are amplified by three tiny and delicate bones: the malleus (hammer), incus (anvil) and stapes (stirrup) of the **middle ear**. The stapes vibrates on the vestibule of the **inner ear**. The inner ear is a fluid-filled, membranous complex of structures that holds the specialized array of nerve cells (the organ of Corti) that respond to the vibrations and convey their reactions through the auditory nerve to our brain. Also associated with the inner ear are the three directionally oriented semicircular canals that are involved in our sense of balance and proprioception of our head (Figure 28-1). All that delicate and sensitive system is enclosed in very hard and thick skull bone.

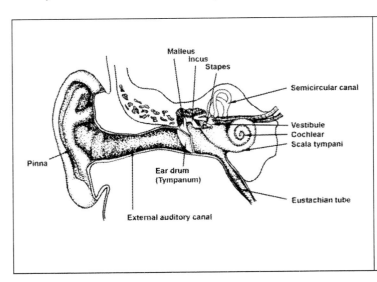

Figure 28-1. The hearing apparatus of a human being. Sound waves, gathered by the outer ear, are funnelled to the middle ear where they are converted to mechanical vibrations by the eardrum and three small bones. The last bone, the stapes, vibrates on the vestibule so its fluid stimulates the sensory cells within the cochlea.

For many years it was thought that honeybees were deaf, even though hearing was known in many other kinds of insects, such as crickets, grasshoppers and cicadas. That has been proven incorrect. Honeybees' ears react to the sound, as do our own, but the mechanism is quite different, and different from that in crickets, grasshoppers and cicadas.   The receptors for air-particle oscillation (sound waves) are in the **Johnston's organs** in the pedicel of the antennae (Figure 28-2).   Johnston's organ in mosquitoes, midges, other flies, and ants are involved in a variety of sensory modalities, including gravity perception, speed of flight, mate recognition by wing vibrations, and hearing.

In the process of hearing in the honeybee, the sound waves deflect and vibrate the antennal flagellum. The intersegmental membrane of the joint between the scapus and pedicel vibrates. That movement stimulates **scolopidia**, a complex of sensory cells, within the pedicel.   The excitation of these sensory cells leads to nervous impulses that are transmitted via the antennal nerve to the bee's brain.

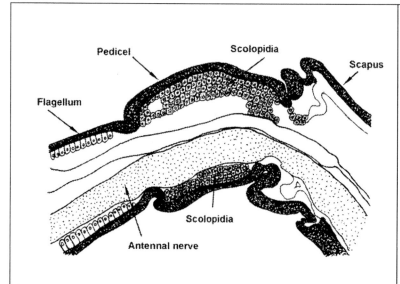

Figure 28-2. Johnston's organ of hearing in the pedicel of the antenna. Sound vibrates the flagellum and pedicel and that stimulates the sensory cells of the scolopidia which, in turn, transmit nervous signals to the brain. Figure 2 – 2 shows the position of the pedicel on the whole antenna.

Bees can hear airborne vibrations at low frequencies—perhaps from about 15 to 500 Hz (one Hertz is one complete vibration per second). Human ears can perceive sounds from 20 to 16,000 Hz or more.   Honeybees use their sense of hearing in Dance Communication (Chapter V – 17).   The high-pitched "piping" noise made by queens when swarming or absconding (Chapter V – 20) are vibrational signals.   The queen "pipes" with her thorax, where the sounds originate from vibrations of the flight muscles, against the comb. She also produces "toots".   The typical pattern of "song" is a 2-second pipe followed by a series of short toots.   The sounds can be carried in the air and on the comb. Attendant workers may stop in their tracks. Other queens in the hive, but still in their cells, may respond by short bursts of sound. The full meaning of the repertoire of sounds is not known.   Other sounds made by honeybees, such as the high-pitched buzzing when they are caught, or the hissing sound made by *Apis dorsata* (Chapter VI – 25) when threatened, may caution would-be threats.

CHAPTER VII – 29

SENSE OF TOUCH & POSITION

The sense of touch is universal among animals. Even *Amoeba* respond to poking and react when they bump into objects as they crawl. In complex animals, such as arthropods and vertebrates, tactile perception is accomplished by special **mechanoreceptive sensilla** (singular is **sensillum**) all over the body, but more concentrated on some parts than others. In insects the tips of the antennae, tips of the mouthparts, feet, and tip of the abdomen are especially sensitive to touch. In honeybees, there are mechanoreceptive sensilla on the compound eyes (see Figure 26-3).

The sense of position of one body part with respect to another (**proprioception**) is crucial to muscular coordination and movement. This form of mechanoreception comes into play through specialized, unicellular stretch receptor neurones within the muscles and connective tissues of the body, and through special paired patches of sensilla on the exoskeleton.

Sense of Touch

Honeybees use their sense of touch on their antennae and mouthparts to determine the thickness and smoothness of the wax walls of the comb. That is important in building new comb and cleaning existing comb (Chapter IV – 15). **Antennation** (touching with the antenna) is probably important in orientation within the nest, on the walls of the cavity, and in following the dances of returning foragers (Chapter V – 17). Touch and taste probably act together in initiating reactions in the bee. Antennae are not called "**feelers**" without reason. Antennation is also important in bees' orientation on flowers. The microscopic level of discrimination by bees gives the cell surfaces of flower petals new meaning as "micro-Braille"! The sense of touch is mediated through hair sensilla (**sensilla trichodea**), as for example, are found on the tips of bees' antennae. Figure 29-1 shows the sensilla trichodea on the tips of the antennae of two kinds of bees, a honeybee and a leafcutter bee in contact with petals of two different species of yellow daisy-like flowers. One can imagine that, as the antenna is moved across the textured epidermal surface of flowers, the sensilla would be deflected in characteristic ways, resulting in stimulation of the nerves beneath the cuticle (Figure 29-2).

Honeybees' capacity to discriminate microtextural cues was demonstrated in an experiment using a small **Y-maze**. Trained honeybees were made to crawl through a one-bee-width tunnel, at the end of which they had a choice of the training texture on one antenna and a new texture on the other. The semi-natural textured surfaces were flower petals, dried and coated with gold-palladium (as for preparation for scanning electron microscopy). Combinations of sunflower petals presented in normal position, upside down, and backward were used, as well as petals from other species. The bees were not fooled and mostly made the choice of texture leading to syrup reward.

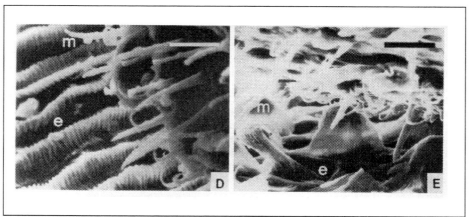

Figure 29-1. Scanning electron micrographs of bees' antennae in contact with the surface of the petals of blossoms from the daisy family (Asteraceae). ((**D**) *Apis mellifera*; (**E**) Megachilidae). Note the correspondence of size and spacing of the mechanoreceptive sensilla trichodea (**m**) with the flowers' epidermal textures (**e**).

Figure 29-2. Diagram of a trichoid sensillum and cuticle of the exoskeleton. As the sensillum is deflected, the nerve within and below (not shown, but see Figure 27-2) is stimulated to send a message to the central nervous system.

Sense of Position (Gravity and Proprioception)

At the joints of the appendages (locomotory, sensory, or feeding), and between the major and secondary segments of the body are complex arrays of sensilla trichodea called **hair plates**. The sensilla of the hair plates between the pedicel and the flagellum of the antenna (a hinge joint with dorsally placed sensilla) and between the head capsule and the scape (a ball joint with a ring of sensilla) of the antenna are deflected as the antenna is moved up and down or from side to side. Stimulation of the nerves in the sensilla provides information on the position of the antenna. The remarkable arrays of hairs on the small projection from the very front of the prothorax and on the front of the prothorax itself sense the position of the bee's head (Figure 29-3)! Thus, the bee can tell if its head is tilted down or up, or turned left or right. The bee can move its head using muscular control, and can allow its head to detect gravity. When a bee is walking on a horizontal surface, the hair plates are in more or less even contact with the back of the head. However, when the bee is oriented vertically upward, the mass of the head with its centre of gravity below the base of the antennae causes its "chin" to fall back so the gap between the top of the head and the front of the **prothorax** widens. When the bee is oriented vertically downward, the "chin" falls forward and the gap between the top of the head and the front of the prothorax narrows. The honeybees' sensitivity to gravitational force through this elaborate set of sensors is crucial to orientation during the **dance communication** for direction (Chapter V – 17).

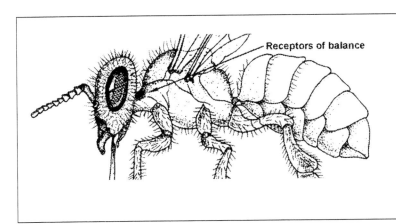

Receptors of balance

Figure 29-3. Diagram of the general position of the main sensor array for orientation to gravity, or sense of balance. Other arrays are present at segmental and limb joints, and on the antennae and mouthparts.

Information from the Nerves

There are two rather different requirements for information reaching the central nervous system depending on whether touch or position are being sensed. For information on position, a continual flow of information is needed, but for touch, information on first contact may be all that is required. The sensilla involved in proprioception typically exhibit **phasic-tonic** nervous responses: deflection results in an initial, rapid nervous activity that continues as long as stimulation continues. Strictly **phasic** responses belong to sensilla that respond immediately to stimulation (deflection) and then cease nervous activity. The sensilla involved in the detection of microtextures are presumably phasic.

Table 29-1. Some Aspects of Neurobiology of Honeybee Sensory Organs

TYPE OF STIMULUS	SENSORY ORGAN(S)	LOCATION	OTHER
Mechanical: Touch	sensilla trichodea	many sites on the body	
Stress/strain	sensilla scolopophora sensilla campaniformia	legs and head lancet of sting	
Gravitational	sensilla trichodea	hair plates on each side of joint between head and prothorax; thorax and abdomen, coxa and trochanter	
Movement	sensilla chaetica	antennae	
Flight speed (in relation to ground)	interneurons	compound eye	
Flight speed (relative to air)	long hairs	head (including compound eyes) and wing	
Energy expenditure (distance)	pressure receptors	wall of honey sac (crop)	
Substrate vibration	sensilla scolopophora	subgenual organ in tibia (foreleg only)	
Heat/temperature (Chapter VII – 30)	sensilla ampullacea, s. coeloconica	last 5 antennal segments	- can detect a change of $0.25^{\circ}C$
Chronosense/Time (Chapter VII – 30)	interneurons in brain	corpora pedunculata	- can register 24-hr cycle
Electrical Sensitivity (Chapter VII – 30)	certain parts of integument	- electrolytic (fluid) parts of body - polarization of the sclerotin layer to which sensory hairs respond	bees upset by thunderstorms, high tension wires
Magnetic Sense (Chapter VII – 30)	magnetite component of granules	fat body	the iron compounds produce a natural magnetic field in the bee's body

CHAPTER VII – 30

OTHER SENSES: HUMIDITY, TEMPERATURE, TIME, MAGNETISM, & ELECTRICITY

Sensing Humidity and Temperature

Honeybees are highly sensitive to **humidity** and to **temperature** (Chapter VII – 27). The importance of such sensitivity resides in the need for the colony to maintain a steady environment and oxygenated atmosphere within the hive for respiration, rearing brood (Chapters V – 17 & 19), and curing honey. They can detect changes in relative humidity of about 5%, and changes in temperature of less than 1°C. The sensors involved are mostly on the antennae and are made up of sensilla ampullaceae and sensilla coeloconica (see Figure 27-1), of which there are about 300 on each antenna. Electrophysiological recordings from the neurones of some of these sensory arrays have shown three kinds: a moisture receptor, and dryness receptor, and a temperature receptor grouped together.

Chronosense or Sense of Time

All organisms seem to have some built-in capacity to measure the passage of time. The natural rhythms of biochemical and metabolic activity translate to the daily rhythm of life. How **circadian** (about one day) **rhythms** are controlled is not well understood, but environmental factors, especially cycles of day and night, are important in setting them. Jet lag would be experienced by any long-distance traveler going east to west or the reverse, not just a human being.

Honeybees have a remarkable sense of time. They have been trained to separate up to nine different tasks over a 9-hour day. That is probably important in foraging because the flowers of different species of plants present their rewards at different times of day. Figure 30-1 illustrates the constrained time of foraging by a number of species of bees on the flowers of one species of plant. What is especially interesting about that plant is that it blooms every second day, and when it blooms its pollen is depleted by the foraging bees within about 1.5 hours. It is unknown if bees can remember multi-day cycles, but certainly they remember when certain forage is available for several days even if the bees are prevented by bad weather from going out.

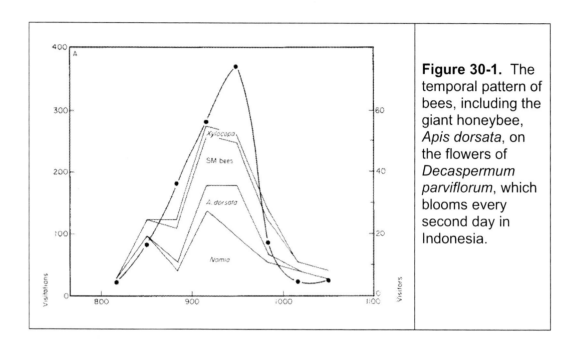

Figure 30-1. The temporal pattern of bees, including the giant honeybee, *Apis dorsata*, on the flowers of *Decaspermum parviflorum*, which blooms every second day in Indonesia.

At a finer level of temporal resolution, it is well known that foraging honeybees compensate in their dance communication for the amount of time that they have been absent from the hive. If a bee is absent from the hive for an hour, a usual sort of trip time, the position of the sun in the sky has changed by 15°. The directional information in the dance communication must accommodate for that, and of course, outgoing foragers responding to that information must adjust their direction, especially if forage is far away (Chapter V – 18). The mechanisms are not known, but it is assumed that interneurones in the brain, especially associated with the mushroom bodies (Chapter III – 9), are involved in the complex computations that are implied by this behaviour.

Magnetic Sense

Although various animals (e.g., migratory birds, pigeons) have been shown to be sensitive to the magnetic field of the Earth, evidence for magnetic sensitivity in bees is less convincing. There is evidence that honeybees respond to magnetism, but their responses do not make much sense. Normal daily fluctuations in the Earth's magnetic field may help in the sense of time.

The mechanism proposed is through the accumulation of crystals of magnetite, or lodestone, in the front of the abdomen during pupation and metamorphosis. Lodestone naturally orients itself to the magnetic field of the earth and was the material of the compasses of ancient human navigation.

Electrical Sensitivity

Beekeepers have noted that honeybee behaviours are influenced by thunderstorms and by proximity of electrical power transmission lines. The general view is that both cause bees to become irritable and defensive. The mechanism is not understood, and in the case of thunderstorms, other environmental cues could come into play. It has been suggested that the cuticular hairs may respond to static electrical charges and electric fields, and that the body fluids may also be influenced.

Section VIII

Practical Beginnings & Application of Bee Biology
Bees and People through the Ages, the Start of Modern Beekeeping, & a Year in a Beekeeper's Shoes

There is ample evidence that human beings have enjoyed the fruits of honeybees' labours since the Stone Age. Beekeeping probably developed from honey hunting and from more systematic honey gathering in prehistoric time. Records of the first hives are as ancient as the Pharaohs of Egypt. Beekeeping did not seem to change much over the millennia. The traditional skeps, log hives, tubes, and pots persisted until the mid-1800s. Then a major change took place. Various beekeepers in Europe and America made advances in developing hives that reduced the damage to the colonies when honey was harvested. In 1851, L. L. Langstroth put the ideas together. He combined the idea of the bee space with "top-bar" frames and invented the modern hive that is used today. That advance revolutionized beekeeping. Inventions such as honey extracting machines, wax comb foundation presses, the "bee escape", the queen excluder, and so on helped make beekeeping a modern, mechanized agricultural activity. Queen rearing, and means of transporting bees reliably and of using them for crop pollination were all technological advances of the age.

Beekeeping today requires almost year-round attention by beekeepers. Spring is a time for careful husbandry and feeding, summer is honey time and pollination time, fall is harvest time, and winter is the time for equipment to be repaired, built, and maintained. From late winter, when queens start laying, to the start of the next winter, the beekeeper's life is dictated by the seasonal cycle of his colonies.

Section VIII – Practical Beginnings and Application of Bee Biology
Chapter 31
Honeybees and Humankind through the Ages – From the Stone Age to the Nobel Prize

CHAPTER VIII – 31

HONEYBEES & HUMANKIND THROUGH THE AGES: FROM THE STONE AGE TO THE NOBEL PRIZE

Beekeeping has been developed in all parts of the world over centuries of accumulated knowledge. Human beings to this day loot wild nests of bees and take the honey. Honey was the only form of sweetener available to people in some parts of the world for thousands of years. Also, for many people honey continues to have cultural, religious and medical significance.

In **honey hunting** and **gathering**, early people may have used methods similar to those used by wild chimpanzees. They poke sticks into subterranean bee (probably stingless bee) nests, withdraw them, and lick them to gather small amounts of honey. Perhaps it is correct to think that even those chimpanzees remember where the sweetness lies and revisit the sites from time to time. Is that honey hunting, or honey gathering? The earliest known evidence of honey gathering relates to *Apis mellifera* in Europe. Some **Mesolithic** rock paintings in eastern Spain date to about 8,000 BP (Before Present) (Figure 31-1).

Several rock paintings in Spanish rock shelters, in South Africa and Zimbabwe and in Asia, depict people at work with bee nests (Figures 31-2 & 31-3). They show honeybee swarms, honey combs, people using ladders to reach the nest, people blowing smoke at bees, etc. The practice of using smoke arose independently in many parts of the world.

Figure 31-1. Honey gathering painted on the wall of a rock shelter at La Arana, eastern Spain, from about 8,000 BP. The figure is feminine and using a bucket.

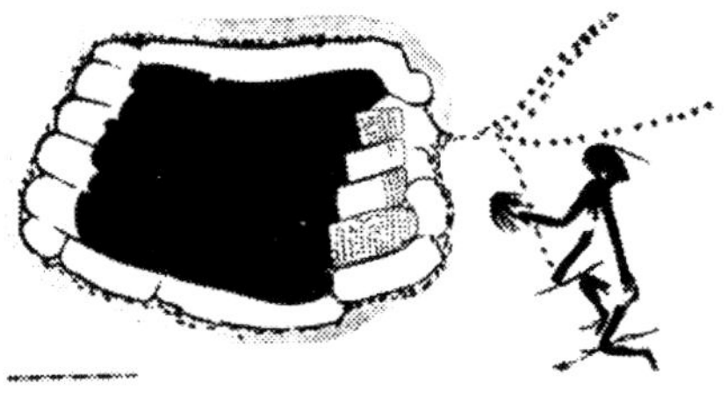

Figure 31-2. Rock painting of honey gatherer smoking a honeybee colony, found in a rock shelter near the Toghwana Dam, Zimbabwe. (Age of painting not known.)

Section VIII – Practical Beginnings and Application of Bee Biology
Chapter 31
Honeybees and Humankind through the Ages – From the Stone Age to the Nobel Prize

Figure 31-3. Post-Mesolithic rock painting showing honey collection from the giant honeybee, *Apis dorsata,* at Rajat Prapat, Central India.

Many of the honey-gathering methods (Figure 31-4), including the use of ropes, bags, baskets, axes and knifes, are used almost unchanged by today's honey gatherers of Africa, South East Asia, India, and Europe.

Figure 31-4. Forest beekeeping in Poland—working a barc (pronounced 'bartch'). The colony is within a naturally occurring hollow tree into which an access door has been cut. The beekeeper can visit the colony periodically and gather honey. Pine trees are thought to be preferred by bees, but oaks and other trees are also used in this method of beekeeping and honey harvesting. Honey was harvested once a year. Trees selected for barcs were cropped at the top to slow their growth and to prevent breakage in wind.

Section VIII – Practical Beginnings and Application of Bee Biology
Chapter 31
Honeybees and Humankind through the Ages – From the Stone Age to the Nobel Prize

Beekeeping presumably began as people learned more about the activities of honeybees and where they lived. Honey gatherers would have appreciated the need to leave some honey behind and to minimize damage to the nest so that they could return and harvest more later in the year. Marking bee trees and other sites where bees nest would have followed to ease finding and perhaps to note ownership by communities or individuals. Presumably particularly inventive people would have recognized the possibility of keeping bees in movable cavities, such as in hollow logs, clay pots, or baskets. They may have understood swarming (Chapter V – 20) and baited their containers with honey and bees' wax to induce colonies to take up residence. At that stage beekeeping was established and the **hive** invented.

A **hive** is a container in which bees are kept. Hives are used for keeping cavity-nesting species of honeybees (*Apis* spp.), stingless bees, and bumblebees. Sometimes the artificial domiciles used for keeping leafcutter bees (*Megachile rotundata*) and orchard bees (*Osmia* spp.) (Section XI) are also called "hives". A hive offers protection for the bees and access for the beekeeper. Hive entrances are designed to be appropriate to the bees' needs to forage, to defend themselves, and to control the environment within. Access to a hive for the beekeeper has taken on many forms, but has become increasingly convenient through biological discoveries applied to hive design.

Some of the oldest records of beekeeping come from **Ancient Egyptian** tomb paintings dated 4,500 BP (Before Present) (Figure 31-5 (B)). The ancient Egyptians used cylindrical clay hives, blocked at one end, with a removable access disc, and at the other end a central entrance and exit hole for the bees. Those beekeepers used smoke blown into the hive to drive the bees to the opposite end. Then the beekeepers could remove almost half the honey.

The selection of a bee as a royal hieroglyph for Upper and Lower Egypt (Figure 31-5 (A)) may have implied that Ancient Egyptians knew of the existence of one special bee, a ruler, in each colony. The Ancient Egyptians used honey in many religious ceremonies. They used wax in cosmetics and mummification, for writing tablets and encaustic painting. Bees' wax was used to seal the wooden coffins of dead and mummified Pharaohs.
It is probable that one of the first parts of the world where **migratory beekeeping** was practiced was on barges or boats on the Nile. They would have followed the spring bloom from Lower Egypt in the south to the Nile Delta of Upper Egypt in the north. Early records of migratory beekeeping also come from Mesopotamia. There, bees were moved from the hills of what is now Kurdistan down to the agricultural areas of the eastern half of the fertile crescent of the valleys of the Tigris and Euphrates rivers.

The earliest written records that relate to keeping bees in hives are from about 1,500 BC. They form part of a Hittite code of laws inscribed on clay tablets.

Section VIII – Practical Beginnings and Application of Bee Biology
Chapter 31
Honeybees and Humankind through the Ages – From the Stone Age to the Nobel Prize

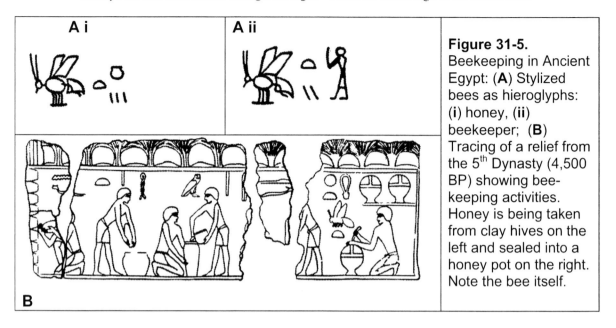

Figure 31-5. Beekeeping in Ancient Egypt: (**A**) Stylized bees as hieroglyphs: (**i**) honey, (**ii**) beekeeper; (**B**) Tracing of a relief from the 5th Dynasty (4,500 BP) showing bee-keeping activities. Honey is being taken from clay hives on the left and sealed into a honey pot on the right. Note the bee itself.

The **Greeks** and **Romans** understood beekeeping. Aristotle (384–322 BC), one of the most important authors of extant Greek writings, wrote that an average hive produced 5–7 kg of honey and an exceptional hive 14 kg. Various other Greek and Roman writers described various aspects of beekeeping and bee biology in accounts of natural history, in prose and poetry. Honey was highly regarded for its value in cooking and to help in health and long life. Honey cakes were thought to have special significance as worthy offerings to the gods. It was used for preserving meat and in producing **mead** (wine made from honey). The hives of the time were roughly cylindrical, arranged horizontally, made of pottery (Figure 31-6), wood, wicker or cork. Records indicate that many Roman farms had apiaries and employed beekeepers.

Figure 31-6. Horizontal terracotta hive in the Greek style. Note the entrance hole. The other end, plugged with a disc, was the beekeeper's access. Similar hives have been found used today in Kashmir for keeping *Apis cerana*.

The transfer of beekeeping technology from Asia Minor and Egypt to Greece and Rome is evident by the similarities between techniques throughout the region. Some of the

Section VIII – Practical Beginnings and Application of Bee Biology
Chapter 31
Honeybees and Humankind through the Ages – From the Stone Age to the Nobel Prize

techniques reached the Indian subcontinent, as is indicated by the similarity of hives used in Kashmir even today to those from the south and west of that period (Figure 31–6). Perhaps the flow of technology was east to west!

Beekeeping in the **Near** and **Middle** East is noted in the Bible and the Koran. Palestine is referred to in the Old Testament many times as a land flowing with milk and honey. In the New Testament it is written that after resurrection Christ asked, "Have ye here any meat?" And they gave him a piece of a broiled fish, and of a honey comb. And he took it, and did eat before them (*Luke* 24: 41–43). In the Koran, Muhammad, the Prophet (AD 571–632) is reported to have said: "Honey is a remedy for every illness, and the Koran is a remedy for all illness of the mind, therefore I recommend to you both remedies, the Koran and honey". In other world religions, reverence for honey is ancient. It is used in Ayurvedic medicine as practiced in Hindu and Buddhist traditions. Bees, especially *Apis dorsata*, are regarded in Asia as portents of good luck and prosperity.

Throughout the Arab world, beekeepers traditionally have used hives of hollow logs of acacia or date palm, or cylindrical clay pots. Nowadays, there is a move to modern techniques using movable frame hives.

Europe is the place of origin for beekeeping as it is generally practiced with *Apis mellifera*. The development of forest beekeeping (see Figure 31-4) seems to have started in the valley of the Volga River and spread westward, reaching the Slavic and Germanic countries at least 2,000 years ago. Part of a tree trunk with a beekeeper's door, dated to about 2,100 BP, has been found in southern Poland. Upright, traditional log hives developed from that practice (Figure 31-7).

Figure 31-7. An apiary of vertical log hives, called "gums", from the 16th century. The beekeeper is **tanging** the bees by banging the stick on the metal basin. This was supposed to make the swarm settle.

Section VIII – Practical Beginnings and Application of Bee Biology
Chapter 31
Honeybees and Humankind through the Ages – From the Stone Age to the Nobel Prize

In other places, people kept bees in **skeps**. Skeps are tallish straw or wicker baskets (Figure 31-8). Today, one often sees the skep used as a traditional emblem on honey labels. The laws of Ireland (instigated by St. Patrick about AD 440) and those of Wales (AD 918) mentioned the management methods of bees and the value of swarms, so it can be assumed that bees and beekeeping were valued in society at the time. In Norse mythology and in Saxon England, mead (honey wine) was a much favoured drink. Beekeeping was part of the life in mediaeval monasteries, and in the countryside. Bees' wax was much in demand for church candles. Beekeeping was well developed in Europe in the Middle Ages. However, it seems that little changed over a thousand or more years. Figure 31-8 could depict beekeeping over most of the period.

Figure 31-8. Collecting a swarm into a skep in the 17th century. The beekeeper's family helps. Note the empty skep lying at the foot of the ladder, the row of skeps, each with two entrances, under the thatched roof and raised off the ground. The child is tanging, and everyone is wearing a veil.

In the Middle Ages, European beekeepers were familiar with various techniques. They had **protective clothing**. Hives were kept under shelter from rain and snow, and off the ground away from damp and mice. **Bee-lining** was understood and used for locating colonies. The technique was to capture bees working at flowers and to confine them in a straw. They were released one at a time, and each bee was followed as far as it could be, then the next one was released and followed. In that way, the home hive would be found. **Drumming** was used to "drive" bees from one container into another. By drumming with one's fingers on the walls of a container containing a colony of bees, one can cause the bees to move away from the drumming. That technique was used to **transfer swarms** and to **combine colonies**. **Expanding the sizes** of the hives was practiced by use of the "eke" (a straw ring that was placed under a skep to extend its volume), and "bell" (a smaller skep placed over a hole made in the larger skep) from management of strong, productive

Section VIII – Practical Beginnings and Application of Bee Biology
Chapter 31
Honeybees and Humankind through the Ages – From the Stone Age to the Nobel Prize

colonies to help reduce swarming and increase yields of hive products. The clever beekeepers of the Middle Ages managed their bees to keep **brood above and honey below** (i.e., the reverse of the natural arrangement). **Smoking** was known from ancient times and 17th and 18th century Britain to obtain honey from skeps; sometimes bees were intentionally killed with fumes of burning sulphur. Skeps with strong colonies were also selected for **overwintering** and providing next year's harvest, and swarms were used to fill empty hives. St. Ambrose was the patron saint of beekeeping and, like the Greek god Trophonius, is depicted with bees and bee hives. Both were associated with safety, welfare, and social responsibility.

The slow advances were aimed at extracting honey and bees' wax while <u>not</u> destroying the hive. Even so, with pots, skeps, and gums, it was difficult not to cause damage. Figure 31-9 shows how the honeycomb was typically arranged in a skep. The comb was cut out of the skep and then processed by putting it in a bag and squeezing the honey out. The honey dripped through the bag and the wax remained within.

Figure 31-9. Looking up into the bottom of a straw skep to see the parallel combs that would be cut to obtain bees' wax and honey.

One of the techniques that allowed for less destructive harvesting of hive products was probably the **top-bar hive**. In those hives, the comb hung from bars that were placed across the top of the hive. The bars could be lifted free and also made a cover over the nest within. The first top-bar hives seem to have been used in Greece (Figure 31-10) in the 17th century.

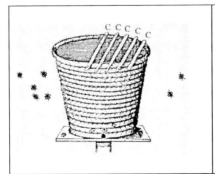

Figure 31-10. Sir George Wheler's drawing (1682) of a Greek top-bar hive.

Section VIII – Practical Beginnings and Application of Bee Biology
Chapter 31
Honeybees and Humankind through the Ages – From the Stone Age to the Nobel Prize

Although the technique attracted interest, little progress was made using it. However, things started to change with the advent of the Industrial Revolution and human curiosity about invention. **Bar hives** and **movable frame hives** became an important part of beekeeping in Europe in the early 19th century. Peter Prokopovitch, a commercial beekeeper in Ukraine, is reported to have kept over a thousand hives and to have been one of the first to use movable frames (Figure 31-11, left). Francois Huber used a book hive to study bee biology (Figure 31-11, right). He could open the hive and read the frames of comb like a book. Glass observation hives had been in use since the 17th century.

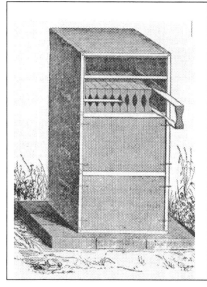 

Figure 31–11. Left is the movable frame hive as used by Prokopovich in the early 19th century. **Right** is the book hive as used by Huber at the end of the 18th century.

Combining top bars and movable frames into a new, and highly practical, hive design came at about the same time as the recognition of the practicality of the "bee space". Jan (H. C. J.) Dzierzon (1811 – 1906) in Poland and Lorenzo Lorraine Langstroth (1810–1895) in the USA opened the way to modern beekeeping by recognizing the value of movable frames. Langstroth revolutionized beekeeping in 1851 with his movable frame top-bar hive (Figure 31-12). Both men studied bee behaviour in depth and used many books on beekeeping from which to draw their inspirations. Both were an experimentalists who tested their ideas on colonies of honeybees. One of the important things Langstroth noticed was that if he positioned the sides and ends of the frames at an appropriate distance (the bee space) from the inner hive walls, the bees would not build comb in the gap nor would they stick the frames to the hive walls with propolis. The frames were, therefore, truly movable.

That distance is the **"bee space"**. For European races of honeybees, the bee space is 3/8" to 5/16" or 6 to 8 mm. If that space is left between all the elements of the hive, between the top bars and the roof, between the top bars themselves, between the end bars and the hive's inside end walls, between the bottom bars in the super above and the top bars of the super below, and between the frames and the side walls of the inside of the

Section VIII – Practical Beginnings and Application of Bee Biology
Chapter 31
Honeybees and Humankind through the Ages – From the Stone Age to the Nobel Prize

hive, the bees leave it empty. Nothing is stuck to anything else, except where internal components rest against each other. There the bees use propolis.

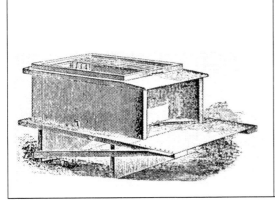 

Figure 31-12. Illustrations of two of Langstroth's original movable frame hives.

Langstroth's hive was in common use in the USA and UK within a decade (Figure 31-13), and within two decades became common in Europe. It is still used today (Figure 13-14).

Figure 31-13. A 19th century British beekeeper using skeps and a Langstroth movable frame hive. The three skeps are in a traditional shelter, off the ground. The beekeeper is inspecting a frame removed from his Langstroth hive as his family looks on.

Although the foregoing discussion on the history of human interactions with honeybees has focused on practical considerations, especially the development of the

Section VIII – Practical Beginnings and Application of Bee Biology
Chapter 31
Honeybees and Humankind through the Ages – From the Stone Age to the Nobel Prize

modern bee hive, apidology (bee science) had made important contributions to beekeeping. Table 31-1 summarizes those discoveries.

Table 31-1. Important Scientific Discoveries in Bee Biology from 1500–1920

Date	Person	Country	Discovery or Contribution
1568	Nikol Jacob		Queens and workers arsie from identical eggs
1586	Luis Mendez de Torres	Spain	The leader of the bee colony is female: A queen who lays the eggs.
1609	Charles Butler	England	Wrote the first book on bee biology, *The Feminine Monarchie,* and noted that drones are male.
1625	Prince Cesi	Italy	The honeybee was the first insect to be drawn by observations made under the microscope.
1630	Francesco Stelluti	Italy	One of the first books on honeybee anatomy, *Descrizzione dell'api.*
1637	Richard Remnant	England	His book *A Discourse or Historie of Bees* noted that workers are female.
1771	Anton Janscha (Janja)	Austria (Slovenian)	Described mating between queen and drone.
1792	François Huber	Switzerland	Noted that workers have undeveloped ovaries but can sometimes lay eggs. He laid the foundations of modern bee science.
1793	Christian Konrad Sprengel	Germany	Wrote the first book on pollination.
1845	Jan (H. C. J.) Dzierzon	Prussia (Polish)	Discovered parthenogenesis in honeybees, and designed movable frame beehive.
1851	Lorenzo Lorraine Langstroth	USA	Developed the modern movable frame hive ("Langstroth hive").
1920	Karl Ritter von Frisch	Germany	Started deciphering the dance communication of honeybees.

Section VIII – Practical Beginnings and Application of Bee Biology
Chapter 31
Honeybees and Humankind through the Ages – From the Stone Age to the Nobel Prize

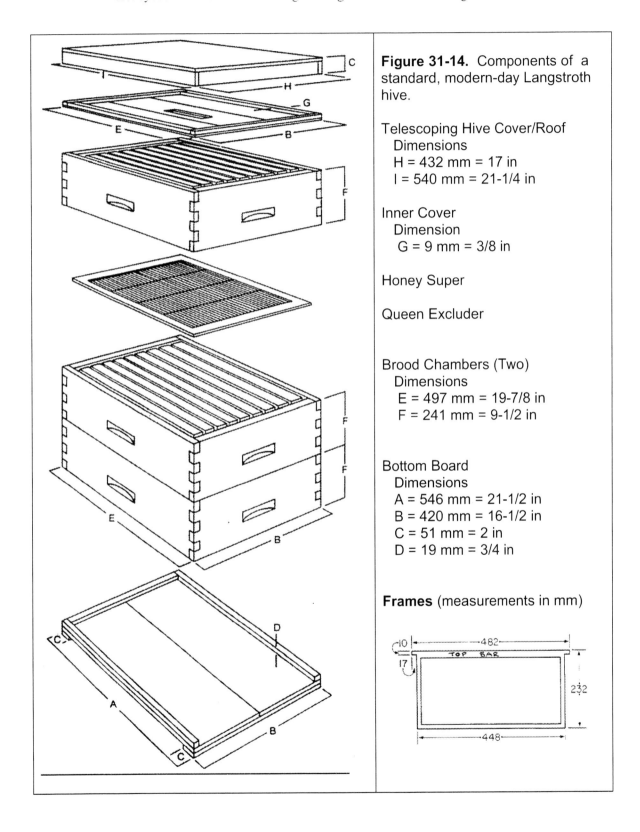

Figure 31-14. Components of a standard, modern-day Langstroth hive.

Telescoping Hive Cover/Roof
Dimensions
H = 432 mm = 17 in
I = 540 mm = 21-1/4 in

Inner Cover
Dimension
G = 9 mm = 3/8 in

Honey Super

Queen Excluder

Brood Chambers (Two)
Dimensions
E = 497 mm = 19-7/8 in
F = 241 mm = 9-1/2 in

Bottom Board
Dimensions
A = 546 mm = 21-1/2 in
B = 420 mm = 16-1/2 in
C = 51 mm = 2 in
D = 19 mm = 3/4 in

Frames (measurements in mm)

Section VIII – Practical Beginnings and Application of Bee Biology
Chapter 31
Honeybees and Humankind through the Ages – From the Stone Age to the Nobel Prize

With the Industrial Revolution came efficiencies in transport and manufacturing. Beekeeping enjoyed a number of technical advances that were made possible by the efficiencies of Langstroth's hive and the amenities of the late 19th century. This period was one of inventiveness, and that atmosphere pervaded into beekeeping. With the capacity to remove and replace frames from the hive with minimal disturbance to the colony, beekeepers soon realized the potential for increasing honey yields while reducing costs. Less labour was needed for every jar of honey harvested, and frames and wax could be recycled back into the hive as complete comb. Some of the important innovations are presented in Table 31-2.

Table 31-2. Important Innovations in Technical Beekeeping

Date	Person	Country	Invention
1851–1852	L. L. Langstroth	USA	Modern movable frame hive
1857	Johannes Mehring	Germany	Comb foundation could be made by pressing bees' wax between embossed rollers with hexagonal designs (Figure 31–16)
1865	Major F. Hruschka	Austria	Centrifugal extractor allowed uncapped frames to be spun so that the honey could be harvested quickly.
1865	Abbé Colin	France	Queen excluder that prevents the queens from ascending into the honey super and laying eggs there. Honey comb from the honey supers has no brood mixed within it.
1888- 1889	G. M. Doolittle	USA	Queen rearing efficiency through thorough understanding of honeybees' life history, especially of queens in larval stage. This innovation resulted in scheduled production of large numbers of queens to serve the expanding industry.
1891	E. C. Porter	USA	Bee escape to prevent bees having descended from the honey supers from getting back, thus keeping honey combs more or less free of bees at harvest time.
1895	M. B. Waite	USA	Migration of honeybees onto crops for pollination services.
1920	H. Dadant	USA	Wired comb foundation added strength for centrifugal extraction of honey.
1926	L. R. Watson	USA	Successful instrumental insemination of queens allowed for highly controlled breeding.
1933–1935	E. J. Dyce	Canada & USA	Creamed, solid honey process that is used today. It also prevents fermentation and improves shelf life.

Section VIII – Practical Beginnings and Application of Bee Biology
Chapter 31
Honeybees and Humankind through the Ages – From the Stone Age to the Nobel Prize

To extract honey from the comb by using a centrifugal extractor, the cappings of the honey comb must first be removed. Various capping knives, combs, and machines have been invented. Centrifugal honey extractors come in two main types, tangential and radial (Figure 31-15). In the tangential extractor, the combs are arranged in holders, or baskets (usually 2, 3 or 4), that are tangential to the circular tub of the extractor. As the apparatus within the tub spins, centrifugal force ejects the honey against the inside wall of the tub. Normally, the frames or their baskets are turned around before all the honey is extracted from one side. That lessens the risk of breaking the comb, but does mean three bouts of spinning for full extraction.

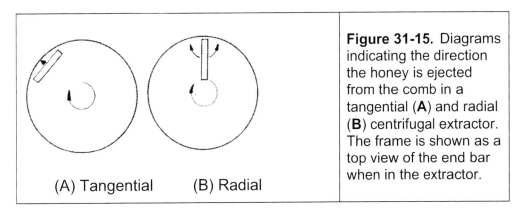

(A) Tangential　　(B) Radial

Figure 31-15. Diagrams indicating the direction the honey is ejected from the comb in a tangential (**A**) and radial (**B**) centrifugal extractor. The frame is shown as a top view of the end bar when in the extractor.

The radial extractor contains a wheel into which the frames are placed like spokes. The top bars are placed outermost. As the wheel rotates, centrifugal force ejects the honey against the inside of the tub. This technique works well for extraction of larger numbers of frames of honey comb and relies on the fact that cells slope slightly down from their entrance to the bottom (Figure 31-16).

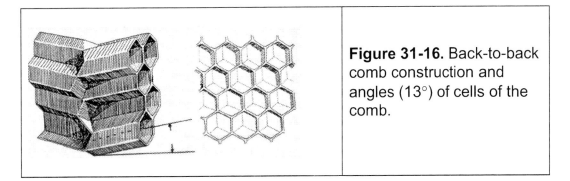

Figure 31-16. Back-to-back comb construction and angles (13°) of cells of the comb.

A number of **bee escape boards** have been designed. All have the same use: to funnel the bees from the upper honey supers down into lower supers and the brood chamber before harvesting the honey. It is useful to have the honey supers as free of bees as possible before taking them to the honey house for honey extraction. E. C. Porter's little invention, a one-way bee valve (Figure 31-17), eased the task of cleaning bees from the honey supers. It is placed in a slot cut to size on a board, like an inner cover (Figure 31-

Section VIII – Practical Beginnings and Application of Bee Biology
Chapter 31
Honeybees and Humankind through the Ages – From the Stone Age to the Nobel Prize

14). The board is placed below the lowest super that the beekeeper wants to remove. Usually the bees in the supers above the escape board have left within a day or two.

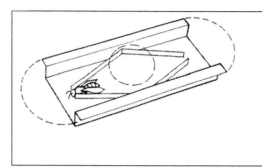

Figure 31-17. A Porter bee escape showing a bee passing between the gentle spring metal gates. Once through, the bee cannot return because the gap is too narrow.

Queen rearing became a well-refined technology through the efforts of G. M. Doolittle. He is considered the father of the modern queen rearing industry. He discovered that priming artificial cups with royal jelly before transferring the day-old worker larvae into them gave the best results in rearing queens. He emphasized that good queens could be reared only by stimulating the swarming urge of the breeder colonies. He took advantage of what was understood about colony reproduction (Chapter V – 20) in his research toward commercial queen rearing (Chapter X – 44). His findings were commercialized in about 1888 and his methods are what are used, with little further modification.

CHAPTER VIII – 32

A YEAR AS A BEEKEEPER

A beekeeper must put effort, knowledge, and judgment into being successful. The main factor to be kept in the beekeeper's mind is to maintain healthy and strong colonies. Those are needed whatever the main aim of beekeeper may be: honey production, production of other hive products, providing pollination services, or breeding bees.

This chapter pertains to beekeeping in temperate climates, such as are found in Canada, much of the USA, and Europe. Its emphasis is Canadian. It presents a very general overview of management throughout the year and is not intended to replace reference to a good "how-to" book, or consultation with an experienced beekeeper. The activities that a beekeeper must follow are dictated by the seasonal and daily rhythms of the bees (Chapter V – 20). The natural, seasonal rhythm of a colony of honeybees (Figures 32-1 & 32-2) can be manipulated to advantage by careful management.

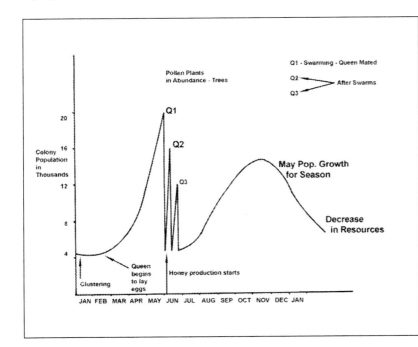

Figure 32-1. (Also appears as Figure 20-3.) The annual cycle in an unmanaged honeybee colony in southern Canada: Q1, Q2, and Q3 represent the issuing of a primary and two secondary swarms. As explained in Chapter V – 20, the losses to colony strength are severe and beekeepers try to avoid this sort of situation through management.

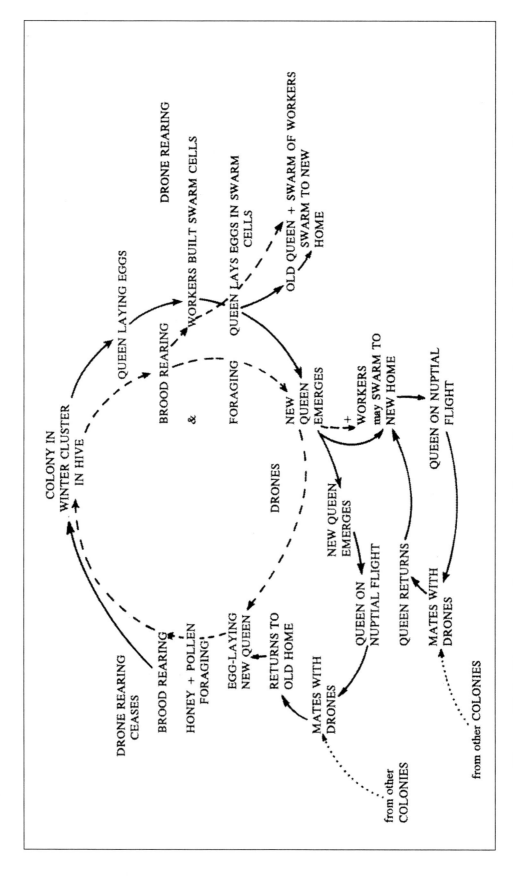

Figure 32-2. Diagrammatic representation of the annual cycle of activities in a colony of honeybees.

Chapter 32
A Year as a Beekeeper

Generally speaking, beekeepers do not want their bees to swarm or abscond. By knowing what causes swarming and absconding (Chapter V – 20), a beekeeper can take appropriate preventive measures. Bees can be encouraged, through management, to gather more honey or pollen, or to be effective pollinators. Breeding bees for queen or colony production is a specialty occupation and requires great skill and attention. This chapter follows the course of the natural rhythm of the seasons, and of activities in a normal, healthy, managed honeybee hive (Figure 32-3).

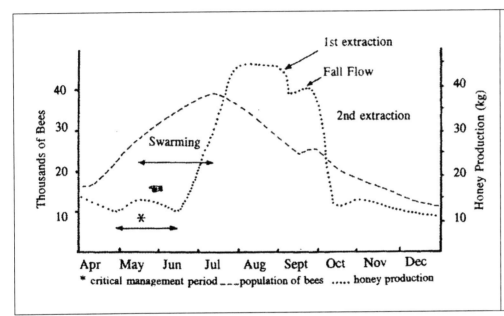

Figure 32-3. (Also appears as Figure 20-4.) The annual cycle in a well-managed honeybee colony in southern Canada. Swarming has been prevented, colony strength is high, and honey production is managed and on schedule.

In the **latter part of winter**, queens have resumed laying eggs and the colony is already gaining strength. However, there are no natural sources of forage available until spring has started. The end of winter and start of spring can be a critical period for the colony and for beekeepers. If winters are particularly long and harsh, cold starvation (Chapters V – 19 & 20) may doom many colonies. Even if the winter is not especially harsh, its duration can cause the colony to starve. Beekeepers must prepare for the worst of winter during the fall before (see below).

Honeybee colonies can be fed in **early spring** (e.g., in March), but that can be done only on mild days when the hive can be opened without risk of chilling the brood. At this time of year, the bees should be disturbed as little as possible. The best time to visit the apiary is between noon and 2 p.m., the warmest part of the day. On a mild day, signs of flight activity at the top entrance of the hive indicate that all is well. On a cool day, beekeepers can tap the front of the hive, or jiggle it, and should hear the "buzz" of an active colony within.

Chapter 32
A Year as a Beekeeper

Whenever it is possible in early spring, the beekeeper should determine the amount of food remaining in the hive. That can be estimated by lifting at the back of the hive. If the hive is not frozen to the ground, hefting it indicates the overall weight of the colony with its remaining honey stores. If the day is particularly warm (over about 12°C), a brief, internal inspection can be made, but that is often a laborious task involving the removal and replacement of the winter packing (Figure 32-10). In general, a colony of honeybees should never have less than 10 kg of feed in the hive. If the colony is alive but light, beekeepers must feed sugar syrup to allow the colony to last for the remainder of the cold weather and until flowers start to bloom. Feeding sugar syrup has a stimulative effect and can lead to enhanced brood rearing. Also, at this time of year, pollen or pollen substitute can be fed to the bees. Providing a boost to the winter-depleted protein stores in the hive has an even greater stimulative effect.

There are various ways of feeding syrup and pollen or pollen substitute to colonies. Syrup is easily fed through a gravity feeder (jar or can with a perforated lid through which the bees can reach the syrup with their tongues) placed in an empty super above the inner cover. Figure 32-4 is of a Boardman feeder, deployed in the entrance of a hive, but the jar and its perforated lid could be used alone.

Figure 32-4. Boardman entrance gravity feeder. This sort of syrup feeder is normally used in the entrance of the hive, but the jar with its perforated lid can be used alone in an insulated, but otherwise empty, super placed above the inner cover. The jar is placed inverted over the hole in the inner cover so that the bees have access to the syrup.

Syrup fed in spring should be made as 1 part sugar to an equal volume of water.

If the beekeeper is unfortunate enough (and in some winters it is all too common) to find hives where the bees have died, the outside entrances must be closed tightly, or the hives removed from the bee yard, so that no robbing can take place once the bees start flying. Robbing seems to stimulate more robbing. Also, it is important to determine the cause of colony death. If disease is indicated (Chapters IX – 33 & 34) then leaving the dead colony in the apiary spreads the disease.

The mild days of the start of spring are busy. Essential tasks are checking the hives for strength, providing feed as needed, removing hives in which the colonies have died, and removing the winter wraps. This is also the time when beekeepers who want to replace colonies should place their orders for packages of bees (if available) or a **nucleus hive** (a small colony of honeybees into which a young healthy queen cell is placed) to arrive as soon as is practical. Availability of packages and nucleus colonies depends on where the bee yard is located. There are restrictions on international shipments of bees into Canada, so obtaining replacement bees early in the year is often not possible and one must wait for domestic supply.

By mid-spring (i.e., April) the bees are often flying freely. They will have taken a cleansing flight to void themselves of the winter's accumulation of faeces in the rectum (Chapter II – 5), and probably will have started foraging on the first flowers. They gather pollen from elms, pollen and nectar from willows, and nectar from dandelions at the start of the season. At this time of year, beekeepers and beekeeping extension personnel often get complaints about "yellow rain" (Chapters V – 19 & VI – 25) on deck furniture, automobiles, and washing hung out to dry.

On a day when bees are flying freely, beekeepers should re-check to see that the colony has the equivalent of **three frames of honey**. If the outer frames are filled with honey the colony is well, but if there is little honey in those frames, the colony should be fed at once with sugar syrup. Again bees can be fed a **pollen** supplement or substitute to further stimulate brood rearing. When the bees can be fed easily in the spring, medicaments should be applied to protect against disease (Chapter IX – 33 & 34). The medicaments can be applied dissolved in syrup, in candy patties, or as dry powder mixed with confectioners' (icing) sugar.

Late spring (i.e., May) is another busy period for the beekeeper and the honeybees. By now, the days are warm enough to allow the beekeeper to make thorough inspections of the colonies. The upper hive entrance needs to be closed, and the bottom board and the lower main entrance must be **cleaned** of debris and dead bees that may have accumulated over winter. On sunny days when many bees are out of the hive foraging, mostly on dandelions and fruit blossoms, the frames in the brood chambers can be inspected to determine the condition of the queen, and to **check** that **the brood** is disease free (Chapter IX – 33 & 34).

If the colony is diseased, beekeepers must (1) assess the situation, (2) take action by administering drugs or control agents, (3) take immediate action if American Foul Brood is present, as prescribed by law, or burn colonies that may be too diseased to medicate and salvage (Chapter IX – 34) the hive and colony, and (4) check other hives carefully. This is the last chance to **medicate** before the start of the first (spring) honey flow. Beekeepers must not medicate within three weeks of the first honey flow to avoid contamination of honey.

If the colony was wintered in one brood chamber, a second brood chamber should be added. If the colony was wintered in two brood chambers, as is the usual practice to provide enough honey stores for the overwintering bees, the **brood chambers should be reversed** (Figure 32-5). Over the winter, the colony has likely migrated within the hive into the upper super (chamber) where provisions were left by the beekeeper. Thus, egg laying and brood rearing has likely re-started there. Reversing the overwintered two-chambered hive puts the brood back at the bottom of the hive where it is wanted. Sometimes, over the winter, the brood becomes split between the two chambers, and the process of reinstating the brood in a single super becomes more complex, involving the separation of the frames in each chamber and reconstitution of the brood area to be as compact as possible in a single chamber.

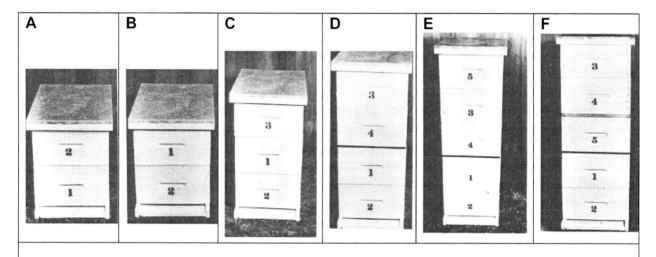

Figure 32-5. Illustration of a series of hive manipulations throughout the season: (**A**) as one would expect to find the colony and hive after the winter, the brood most likely would be in super #2; (**B**) with the brood chamber supers reversed to place the brood area in the bottom super; (**C**) the first honey super #3 is added; (**D**) the second honey super #4 is added, but super #3 is placed above it to encourage the bees to fill #4; (**E**) super #5 is added (one could continue the policy shown in D, but manual labour is a factor); (**F**) the escape board is added to allow extraction of honey from supers #4 & 3, and #5 is placed so that the bees will fill it after the filled supers and the bee escape are removed and replaced—if need be, with an empty super for a continuing honey flow.

From late May to early July, some colonies attempt to swarm (see Figures 32-1 & 32-2). Beekeepers watch for signs of crowding (Figure 32-6), lack of empty cells in which the queen can lay, spots of white wax on the tops of frames, and the presence of swarm cells at the bottom of the brood chamber (Chapter V – 20).

Figure 32-6. A clear sign of an overcrowded hive in early summer.

Supers are added to the hive to alleviate the crowding and whenever one is needed during the summer honey flows.

Early summer (e.g., in June) is the time to look around the countryside and see what honey plants are in bloom. By now, clovers should be in bloom with a wide variety of other nectariferous plants that initiate the summer honey flow. The honey plants vary from place to place, but floral calendars that indicate the general flowering phenology in a region are often available for beekeepers. There are several books, relevant to various parts of the world, on "honey plants". At about this time, beekeepers check that the queen is in the lower two chambers and laying well. The brood pattern should be checked. One good pattern is that of more or less uniformly aged immature bees (eggs, larvae, pupae in sealed brood comb) in large areas of comb. A **queen excluder** (Figures 31-14 & 32-7) may be placed atop the second brood chamber to prevent the queen from getting into the honey supers above. Now is probably the time to add the first honey super to the hive. As summer progresses, more honey supers are added as needed.

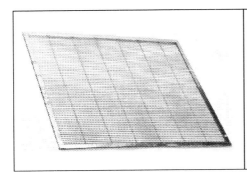

Figure 32-7. A metal frame queen excluder. The metal wires are placed so close together that worker honeybees can pass through, but the queen and drones cannot.

If the beekeeper has frames of foundation (Figure 32-8) that need to be used on the hive, they should be added early in summer. The bees expend a lot of energy in making bees' wax to build comb. The process is called "drawing" the foundation into comb, but really the bees do not draw out the foundation—they build wax onto the hexagonal pattern to make the cells. Frames of foundation are best placed in the upper brood chamber, but can be placed in honey supers. It is not advised to place frames of foundation directly above the queen excluder. Depending on the beekeeper's aims for keeping bees, the queen excluder may or may not be used. For hives for pollination, it is not needed. Such hives are usually two supers deep only, in any case. Larger hives are difficult to move from place to place.

Figure 32-8. A frame of foundation. Note that the supportive wires are visible and the frame is full of honey ready to extract.

As the summer progresses, colony strength should rise to about 40,000 worker bees (a strong colony) by the time the first major honey flow starts. During this period, the brood chamber of each colony should be examined at 10-day intervals for queen or "swarm" cells (Chapter V – 20). Also, beekeepers should watch for obvious signs of overcrowding (see Figure 32-6) and take action by adding a super.

Queen cells strongly indicate that the colony intends to swarm. A diligent beekeeper would destroy the queen cells during inspection, and carry out other management to reduce the likelihood of swarming. It must be remembered that honeybees also build queen cells when their own colony queen is failing for some reason. Keeping a record on each hive is useful because then the age of the resident queen can be looked up; if she is older than two years, requeening should be considered seriously.

To **requeen** the colony, the beekeeper should obtain a young, mated queen from a reputable bee breeder. The new queen is delivered in a small, screened-in cage (e.g, a Benton cage, Figure 32-9) that contains a few accompanying workers and has an exit hole plugged with sugar candy. Requeening involves finding the old queen and killing her, then introducing the Benton cage between two frames in the middle of the brood chamber.

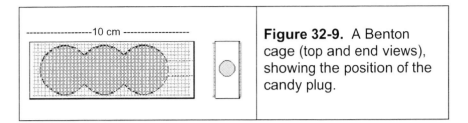

Figure 32-9. A Benton cage (top and end views), showing the position of the candy plug.

The resident bees, and the queen's company, chew the candy plugs so the new queen can escape into the hive. The process usually takes a day or so, and during that time, the hive's workers come to accept the new queen's pheromonal signals. To just introduce a queen directly usually results in the resident workers' forming a ball of bees ("balling") around the new queen and killing her. After a few days, the Benton cage is removed and the colony checked that all is well with the new queen. She should begin to lay within about 10 days.

A less efficient, but more natural, system is to kill the old queen and destroy all but one of the largest queen cells. The bees then raise a new queen of their own. Beekeepers should examine the colony about three weeks later to see that the new queen is laying. Remember, a virgin queen must mature and mate before she can lay.

By **mid-summer, July** and **August**, the colony should be in full swing in foraging and laying in honey. Beekeepers continue to check the colony for queen cells in case they need to prevent **swarming**, but as the hive increases in height as honey supers are added, this become more and more impractical. Some beekeepers may remove honey as the season progresses. They may want to harvest "floral-source honey" from a particular plant that is in abundance around the hives, or they may want to harvest a "seasonal blend honey" that reflects the floral sources at a particular time of year (Chapter X – 40). The beekeeper may wish to use a **bee escape board** to empty the honey supers of bees (Figure 31-17). As honey supers are removed, empty ones are added, until the final honey harvest. Some beekeepers simply add supers to their hives all summer long, and then harvest all their honey at one time.

In **early fall, September** colonies of honeybees have started to wane in strength. The fall flowers, asters and goldenrod, often offer a strong honey flow, so beekeepers need to watch the flowers and their colonies so they do not miss the late season bonanza. Once the fall flowering has ceased, usually after the first harsh frost, beekeepers recognize the close of the season is nigh. On warm autumn days, beekeepers **harvest honey** for the final time of that year. Beekeepers should remove the queen excluder before replacing the super of honey and check for a laying queen.

Harvesting the honey involves **removing the honey supers from the hives** and transporting them to the location of the extractor (the honey house for commercial beekeepers, or the kitchen for hobbyists). The frames of honey must be **uncapped**. There are various tools designed for that, including knives, scrapers, planes, and larger uncapping machines. The uncapped frames, with their comb dripping with honey, are **placed in the extractor**, spun to centrifugally remove the honey. The empty combs are conveniently stored in the supers from which they came, and can be put away for future use. The honey in the extractor is removed by various means, including pumps and pipes in commercial operations, but is **filtered** (cloth screen is all that is needed) to clean it of wax particles and other solid material and **allowed to stand** while bubbles of air rise. Wax particles and air give honey a cloudy appearance which is not desirable (Chapter IX – 40). The honey can then be bottled for sale, or placed in bulk containers such as drums and tanks.

But this is the time to prepare the hives for winter. If a colony is strong, and the hive is heavy with honey and pollen in the two standard supers of the brood chamber, it may be left that way. If the colony is strong, but the hive light, all that may be needed is to place some frames of honey from above into the brood chamber. Using honey from other hives can be risky: it may spread disease. If the colony is not strong and has little honey in the brood chamber, it can be combined with another colony.

In fall, usually in mid- to late October, beekeepers make their final checks of their colonies. Each hive colony should go into winter with about **30 kg of honey.** Remember that the bees are active all winter long: they do not hibernate. The final weight of hive for overwintering should be about 60 kg, broken down as follows: **Honey = 25–30 kg, Wooden ware = 25 kg, Wax = 5 kg, Bees = 2–3 kg**. At this time, **sugar syrup** can be fed to bring the colony up to winter weight. The same, or different, methods as used in spring can be used. A heavier syrup, two parts sugar to one part water, than is used in spring is preferable for fall feeding. **Pollen** stores should also be obvious in the hive as the colony is prepared for winter. Supplemental pollen or pollen substitute can be provided at this time. The bees store it away for use when brood rearing re-starts in late winter. Beekeepers often **feed drugs** in the fall, especially to guard against *Nosema* (Chapter IX – 34).

Because honeybees form a cluster and stay warm during winter, hives should not be closed up tight. The hive needs to be able to vent off stale air and excess moisture from respiration, and take in fresh air. Also, during particularly mild weather in winter, the bees may take the opportunity for a cleansing flight. Beekeepers use entrance reducers to help keep out cold air, and to prevent mice from entering the hives. They also provide a small, top entrance/exit hole for ventilation and for bees to come and go on warm days when snow is deep and blocks the main entrance. Hives should be provided with **winter insulation** to reduce heat loss. Usually, an insulating inner cover is placed over the top super and a tar-paper or plastic wrap over the entire hive (Figure

32-10). There are various strategies for overwintering honeybees outside; in especially cold parts of the world, indoor facilities are commonly used.

Figure 32-10. Overwintering bees with winter wraps in place.

During the winter months there is relatively little external sign of activity that a colony of honeybees is alive inside the hive. The queen lays no, or few, eggs from about October to mid-January. But before the end of January, the queen will re-start **laying**. Then the energy and nutritional demands of the colony rise sharply, and the stores laid in the year before are rapidly consumed (Chapters V – 19 & 20).

Once the bees are safely accommodated for the winter, and equipment is all stored, beekeepers prepare for the next year. This is the time for repairing equipment, making new hive components, repairing or purchasing beekeeping paraphernalia (Figure 32-11) and, with luck, taking a well-earned rest.

 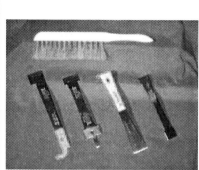 

Figure 32-11. Beekeeping paraphernalia. The beekeeper (**left**) is dressed in protective clothing, including a hat and veil. His trouser legs are tied around the ankles to prevent bees from crawling inside. Beekeepers often wear protective gauntlets (gloves), but they restrict dexterity. In the beekeeper's right hand is a hive tool (the **centre** panel shows several types of hive tool and brush for sweeping bees off comb for inspection). On the front of the hive is the beekeeper's smoker. The **right** panel shows the smoker's workings as a bellows that pumps air into the combustion chamber. The usual fuels that beekeepers use are burlap, wood shavings, leaves, cardboard, and so on. The fuel is ignited in the combustion chamber. If it is dry and highly flammable, green grass can be added to cool the fire and produce more smoke. The lid is closed once the fuel is ignited. By pumping on the bellows, the beekeeper can keep the smoker smoking. The smoke drives the bees away, and seems to calm them, too. When opening a hive, the beekeeper first would make a few puffs of smoke in the entrance. Smoke would be puffed on the bees crawling on the top bars of the frames to drive them down amongst the frames. The hive tool is used to pry the frames up from the lip in the end walls of the supers, to separate the frames, and to help lift the frames out of the super until the beekeeper can grasp the lugs at the end of the top bar. The bees glue the frames in place with propolis (Chapter IX – 42) at points of contact with the end walls of the supers and each other.

Section IX

Problems in the Bee Yard: Beekeepers' Headaches
Diseases, Parasites, Pests, Predators, & Poisons

As with any animal, honeybees suffer from the **Five P**s: **Pathogens** that cause diseases, **Parasites**, **Pests**, **Predators**, and **Poisons**.

Viral, bacterial, and fungal pathogens attack either adult or larval honeybees and cause serious diseases. There are also several parasites that cause diseases of adult or larval honeybees, or both. Bees do have natural systems of defense. They can build up physiological immunity to diseases (Chapter II – 6). They can also adapt behaviourally through engaging in hygienic activities when disease is present in the colony. They groom each other, and remove corpses from the hive. Antibiotics, chemicals, and plant products can be used to reduce losses to the colonies and the beekeepers. Genetic selection for disease resistance and hygienic behaviour has also proven to be a useful tool in the fight against bee diseases (Chapter VI – 23). By combining those approaches with sanitary management by beekeepers, **Integrated Pest Management** (**IPM**) can go far toward assuring the health of honeybees in beekeepers' care.

Pests and predators are mostly nuisances in beekeeping, but usually do cause serious losses. There are techniques for reducing the risks of their attacks whether they be from other insects, frogs, toads, lizards, birds, and various mammals, including bears with their notorious love of honey.

Honeybee poisonings by pesticide applications in agricultural settings and forest management are all too common. From time to time and place to place, they are a major cause of beekeepers' losses and premature aging of beekeepers themselves! Avoiding losses of honeybees caused by pesticide use requires co-operation between beekeepers and pesticide users.

Section IX – Problems in the Bee Yard: Beekeepers' Headaches
Chapter 33
Diseases of the Brood

CHAPTER IX – 33

DISEASES OF THE BROOD

The most serious of diseases for larval honeybees is **American Foul Brood (AFB)**. Its name "American" does not indicate its origin, only that the causative agent, the pathogen, was first elucidated in the USA. It is a bacterial disease, caused by *Bacillus larvae*. The spores of this bacterium are highly resistant and can survive for years. The spores are dispersed by adult honeybees, in honey and bees' wax, and on hive tools and equipment. It spreads rapidly, is difficult to control, and, if unchecked, destroys colonies. Sanitary beekeeping is a must to reduce risk of serious infection by this bacterium. Some control can be achieved by antibiotics. Oxytetracycline HCl (Terramycin ®) is usually administered to colonies in their hives, either with syrup or dry and mixed with icing sugar, in spring and fall as a preventive (prophylactic) measure.

Hives infected with AFB smell bad, as the name suggests. Infected brood has sunken, greasy, and perforated cappings (Figure 33-1). The brood pattern is often uneven. Within the cells, the dead pupae tend to have their developing mouthparts stuck to the top of the cell. Dead larvae and pupae rot, lying on their backs, in their cells, and their corpses are gluey and stick to a toothpick, or piece of grass, thrust into the cell. The "ropiness" test distinguishes AFB from European Foul Brood. After a few days the corpses dry and form scales on the bottom of the cells. Those scales are hard to remove, and are a reservoir of bacteria.

Because of its virulence, AFB is a reportable disease in many places. That means that an official notification must be made to the apiarist. Apiary inspectors visit the diseased bee yards and inspect the colonies. They are empowered to burn diseased colonies in an attempt to bring the disease under control.

Recently, antibiotic-resistant AFB has been reported from various places. It constitutes a serious threat to beekeeping if it is not controlled.

Section IX – Problems in the Bee Yard: Beekeepers' Headaches
Chapter **33**
Diseases of the Brood

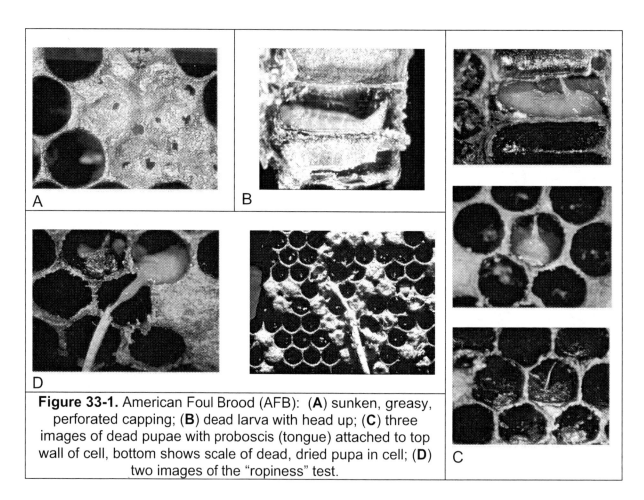

Figure 33-1. American Foul Brood (AFB): (**A**) sunken, greasy, perforated capping; (**B**) dead larva with head up; (**C**) three images of dead pupae with proboscis (tongue) attached to top wall of cell, bottom shows scale of dead, dried pupa in cell; (**D**) two images of the "ropiness" test.

European Foul Brood (**EFB**) is a less serious problem, but is common. Its symptoms are similar to those of AFB: irregular brood pattern; sunken, discoloured and greasy-looking cappings; and a nasty smell. Inside the cells, larvae dead of EFB are twisted, with their heads raised (Figure 33-2). The scales of the dried corpses tend to be rubbery and can be more easily removed from the cells than can those of AFB-killed larvae and pupae. The "ropiness" test does not produce the stringy material adhering to the toothpick, rather, the corpse leaves a moist residue. It is caused by another bacterium, *Melissococcus pluton*. The disease usually makes its presence known in mid- to late spring. It is quickly transmitted throughout an infected colony, and can be moved from colony to colony by bees, and on contaminated tools and equipment. It rarely kills colonies directly, but weakens them, making them susceptible to other stresses.

Oxytetracycline HCl (Terramycin ®) is usually administered, when needed, as a treatment for EFB-affected colonies in their hives, either with syrup or dry and mixed with icing sugar. The prophylactic treatment of colonies to guard against AFB also guards against EFB.

Section IX – Problems in the Bee Yard: Beekeepers' Headaches
Chapter 33
Diseases of the Brood

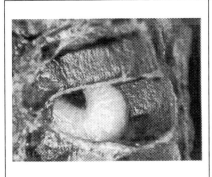
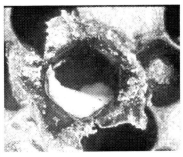
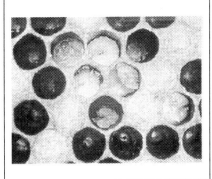

Figure 33-2. European Foul Brood (EFB) showing twisted position of dying and dead larvae.

Powdery scale disease is rare, but it too is caused by bacteria. Sometimes, larval honeybees may be infected with more than one microorganism. It is not clear that all such infections are pathogenic. Some may be secondary infections by organisms that would otherwise be harmless.

The most important viral disease of larval honeybees is **sacbrood**. As the name suggests, infected larvae become sac-like (Figure 33-3). They can be removed from the comb easily with a fine pair of forceps. Dead larvae have darkened heads. The corpses tend to be scattered through the brood pattern. It is not regarded as a serious disease in colonies of *Apis mellifera*. It usually occurs in spring to early summer.

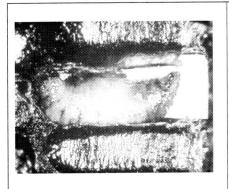
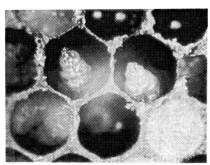
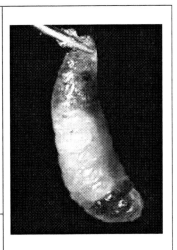

Figure 33-3. Sacbrood showing darkened, raised head of dead larvae and sac-like larva removed from cell.

The equivalent disease of *A. cerana* created havoc with beekeeping and wild colonies over a period of about 10 years. It was detected in Thailand in 1976, and had spread north and west into eastern India by 1979. By 1984 and 1985, it had swept as a plague (epizootic) all the way through Nepal, northwestern India, into Kashmir and

Section IX – Problems in the Bee Yard: Beekeepers' Headaches
Chapter **33**
Diseases of the Brood

Pakistan. Over that period it has been estimated that 80–90% of all colonies of *A. cerana* perished. Since then, populations have recovered throughout the range. There is no cure for this disease, and why the epizootic occurred is not known.

Fungi often cause diseases in insects. The genus *Ascosphaera* attacks many insects including honeybees, bumblebees, leafcutter bees, and orchard bees. *Ascosphaera apis* infects larval honeybees, causing **chalkbrood**. The gross symptoms are hardened mummies of infected larvae (Figure 33-4). They look like small pieces of white chalk, or dirty chalk in the cells. Infection of the larvae occurs about 3–4 days after hatching from the egg. The fungus takes over the entire body of the larva. This fungus, like many others, has two mating strains (+ and –), so when both are present in the same mummy, the fungus spore cyst can form. That is responsible for the grey, dirty, appearance.

There is no curative agent for chalkbrood. Normally, a strong colony of honeybees will eliminate the disease on its own.

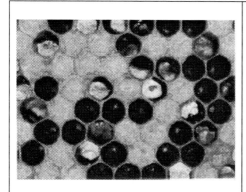 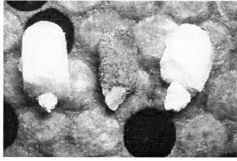

Figure 33-4. Chalkbrood mummies in their cells, and some removed to show colour variation.

Stonebrood is also a fungus ailment of larval and adult honeybees. The causal agent, *Aspergillus flavus*, is common in the soil, and can even infect birds and mammals. Stonebrood mummies are yellowy-green.

The remaining serious disease of larval honeybees is caused by ***Varroa* mites**. These parasitic mites infect larvae, pupae and adults. These are covered under diseases of adult bees (Chapter IX – 34).

CHAPTER IX – 34

DISEASES OF ADULT HONEYBEES

Microscopic examinations are necessary for the diagnosis of most diseases and parasites in adult honeybees. The only one visible to the naked eye is **varroatosis**. To cover the requirements for dissection and examination of bees, two microscopes are required: the stereoscopic low power model and a high power compound microscope. The bees are usually pinned onto a cork for dissection. The body parts are examined carefully and the necessary specimens are taken for further detailed studies.

A number of **viruses** have been identified and found to cause diseases in the adult honeybees (Table 34-1). The sacbrood virus and black queen-cell virus infect immature bees. Viral infections are widespread, and cannot be cured by drugs or other treatment.

Nosema disease is one of the most widespread of adult bee diseases. It is caused by the pathogen *Nosema apis*, a protozoan (Microsporidia) that develops within the epithelial cells of the ventriculus (Chapter II – 5) midgut of adult honeybees. The spores (Figure 34-1) are transmitted in feces from infected bees, usually when young bees clean contaminated combs. The spores are ingested, pass through the oesophagus, honey stomach and proventriculus. When they enter the ventriculus, they germinate through a pore and germination tube (polar filament) which allows the protozoan cells to invade individual epithelial cells of the ventricular lining. There, the protozoan nucleus goes through many divisions, resulting in thousands of new spores. The infected epithelial cells are shed into the ventriculus and burst. Hence, the whole of the intestinal content becomes infective once defecation has occurred.

The gross symptoms are crawling bees with distended abdomens and dislocated wings. The bees are unable to sting. In some respects, the symptoms are similar to some viral infections (below). The infection weakens and kills the bees, and causes the hypopharyngeal glands to shrink and cease producing royal jelly. Colonies with Nosema infections weaken and can die. Often, the diseased bees crawl onto the front of the hive to defecate, hence the common name "bee dysentery". Soiling the front of the hive with infective faeces spreads the disease even more.

Nosema disease is treated with fumagillin (Fumidil-B®). Beekeepers often add this medicine to syrup for fall feeding. The disease tends to be most prevalent in spring.

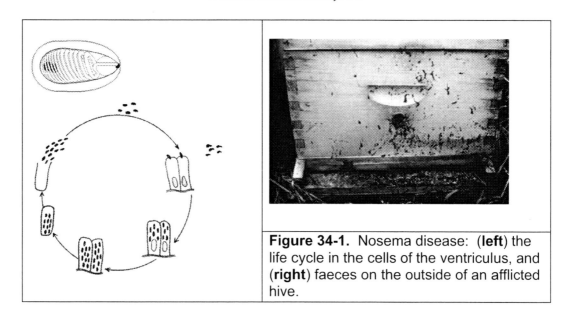

Figure 34-1. Nosema disease: (**left**) the life cycle in the cells of the ventriculus, and (**right**) faeces on the outside of an afflicted hive.

There are other microbial and viral infections of adult bees, but they are generally uncommon. Some appear to be innocuous, but others cause disease. There are no treatments for them (Table 34-1).

Table 34-1. Other Less Important Infections Recorded from Adult Honeybees

Infection	Causal agent	Notes
Septicemia	*Pseudomonas aeruginosa* (Bacterium)	The bacteria attach the bees' connective tissue. They literally fall apart when handled and smell putrid.
Amoeba	*Malpighamoeba mellificae* (Protozoa)	Infects the Malpighian tubules, which swell. The disease weakens the bees and may be called "dwindling" or "disappearing" disease.
Gregarines	Gregarinidae (Protozoa)	Four species of these large Protozoa are known to affect honeybees. They do not appear to be detrimental.
Flagellates	*Crithidia* sp. (Protozoa)	They are found in the alimentary tract and appear to be harmless. *Crithidia bombi* seems to be virulent in some species of bumblebee.
Chronic Bee Paralysis	CBPV (Virus)	Causes the bees to become listless, trembling crawlers. The bees appear hairless, shiny and greasy. Their wings may be disjointed and their abdomens swollen.
Acute Bee Paralysis	ABPV (Virus)	Afflicted bees die rapidly, but the virus is not often encountered.
Filamentous Virus Disease	Nucleocapsid (Virus)	Turns the haemolymph milky.
Other Virus Diseases		Mostly associated with other infections.

Chapter 34
Diseases of Adult Honeybees

The parasitic bee mites have wreaked havoc on the beekeeping industry. **Acarine disease (acariosis)** is caused by infection of adult honeybees by the **tracheal mite**, *Acarapis woodi*. Although it was known from Europe, and is probably indigenous to honeybees there, it did not make its way to other parts of the world with the early migrations of honeybees. When it came into North America, it spread rapidly with devastating consequences. **Varroatosis** is naturally a disease of *Apis cerana* in Asia. It is caused by parasitic mites of the *Varroa jacobsoni* complex, of which *V. destructor* is the main species affecting *A. mellifera*. Somehow, the mites transferred to *A. mellifera* and they have invaded Europe and North America, again with devastating consequences.

The **tracheal mite** infects the large tracheae (breathing tubes) of the thorax (Chapters I – 3 & III – 10) (Figure 34-2). There the mites reproduce and can proliferate enough to more or less block those tracheae. The mites feed on the bee's haemolymph by piercing the tracheal walls with their stylet-like mouthparts. Apart from the weakening of the bees caused by that feeding, the presence of the mites in those important tracheae can severely limit an infected bee's capacity to breathe. That, in turn, has serious repercussions on an infected colony's ability to respire and thermoregulate (Chapter V – 19) and probably explains why winter mortality is so high among infected colonies.

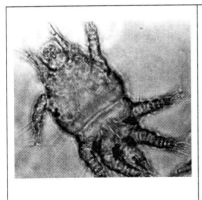 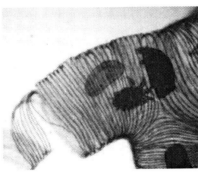

Figure 34-2. Tracheal mites: note the piercing mouthparts (**lower right of left image**) and the infestation of several life stages in a trachea dissected from an infected honeybee (**right**).

The mites complete their life cycle in about 28 days (Figure 34-3). They infest young bees, but at maturity and toward the end of the life of the host bee, they crawl out of the infested trachea and, attaching themselves to passing bees, enter their tracheae. All the four stages (egg, larva, nymph, adult) of the mite may be present at one time in the same trachea (see Figure 34-2, right).

Diagnosis of acariosis requires dissection of bees. They are pickled in 70% alcohol (ethanol), placed on a convenient cutting surface on their backs, and "sliced" so that the prothorax is removed and the tracheae exposed. The sliced part, without the head, is examined under a microscope, and the mites counted.

Tracheal mites have been known in Europe for many years, and presumably are native there. For unknown reasons, the mite did not become established in North, South and Central America, Australia, or New Zealand when honeybees were first transported around the world from Europe. It spread from Mexico into the USA between 1980 and 1984. It was known from Brazil earlier. Once the mite entered the USA it spread rapidly, causing havoc with the beekeeping industry. The plague of tracheal mites in the USA was the reason for the closing the Canada–USA border to honeybee importation to Canada in 1988.

Although acariosis has taken a back seat to varroatosis (see below), it may still be quite a serious cause of overwintering losses of honeybee colonies. Treatment of hives by fumigation cuts back mite populations. Fumigants such as menthol, thymol, camphor, and formic acid have been found effective and have become available in commercial treatments such as Apiguard ® and Miteaway ®.

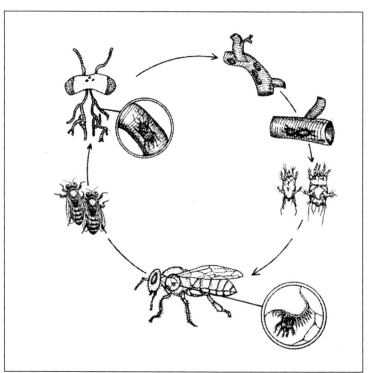

Figure 34-3. The life cycle of the tracheal mite, *Acarapis woodi*. At the bottom, note the mite transferring into the spiracle of its next victim.

Varroa destructor is an external parasitic (ectoparasitic) mite of honeybees (Figure 34-4). *Varroa* mites were first discovered associated with *Apis cerana* in Java in 1904. The mite originally described was given the scientific name *Varroa jacobsoni*. Since then, scientific investigations have revealed a number of species of closely related *Varroa* mites. That which has become associated with *A. mellifera* is *V. destructor*. It made its way into Europe from the Far East. It became widespread in

Europe, causing great destruction. It entered the USA, probably in the early 1980s. It was first reported to be widespread in the USA in 1987 and 1988, and is presently probably the worst disease of honeybees to affect beekeeping.

Mature mites of *Varroa destructor* are reddish brown and large and easily seen with the naked eye on white larvae or pupae. They are more difficult to see on adult bees because they are similar in colour and hide in the intersegmental folds. They cause damage by feeding on the haemolymph (Chapter II – 6) of larval, pupal and adult honeybees of all castes.

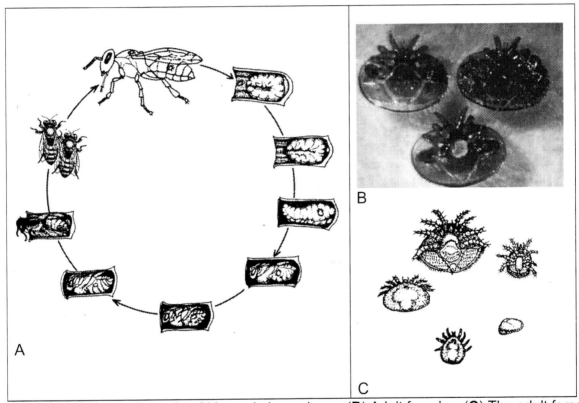

Figure 34-4. (**A**) The life cycle of *Varroa* in honeybees; (**B**) Adult females; (**C**) The adult female (**top**), adult male, egg, protonymph, and deutonymph are shown in clockwise order.

Although drone larvae are preferred as hosts by *Varroa* females, smaller worker larvae support most mite reproduction in infected colonies (Figure 34-5). The adult female mite lays four eggs in cells occupied by week-old brood. More than one adult female mite may use a single larva as its host. She lays eggs over a four-day period starting about 2.5 days after she has entered the cell and taken up residence beside the larva. Immature male mites take 5–6 days to pass through the stages (egg, protonymph, deutonymph (Figure 34-4)) to adulthood; the larger female mites take two days longer. Thus, at 21 days, when the adult bee is uncapped and ready to emerge, so are the adult mites.

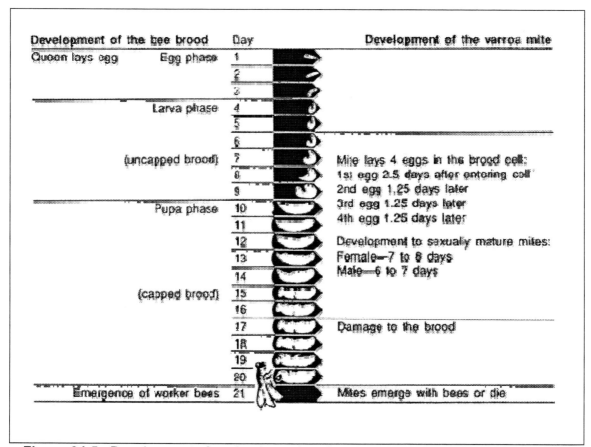

Figure 34-5. Development of worker honeybee from egg to adult and the co-incidental development of *Varroa destructor*.

The mites feed on various parts of the bees' bodies by inserting the stylet-like mouthparts through the cuticle and feeding on the haemolymph. Feeding on the developing wing bud area of bees may result in deformed and nonfunctional wings (Figure 34-6).

The mite's reproductive rate is low (4 per female) in the honeybee colony, so it takes several years for the mite population to rise high enough to seriously weaken the bee colony. It is difficult to detect in the initial stages of infection when there are few mites in the hive. However, once the infection has taken hold, visual inspection of adult bees (a sample of 500–1,000 bees can be collected and killed as for sampling for acariosis), of capped cells (especially of drones), and of hive debris on the bottom board reveals their presence. For monitoring, the quick test is useful. It involves freezing (with liquid nitrogen) an area of brood and cutting that from the comb for return to a laboratory for inspection. Some of the same fumigation techniques as used for control of tracheal mites can be used. Apistan® and Checkmite+®, both **miticide**-impregnated strips, are used to kill *Varroa* within the hive (Figure 34-7). Because miticides leave residues in

x and honey, and because *Varroa* is difficult to control and has developed resistance
iiticides in some places, the search continues for effective protection (see also small
beetle in Chapter IX – 35).

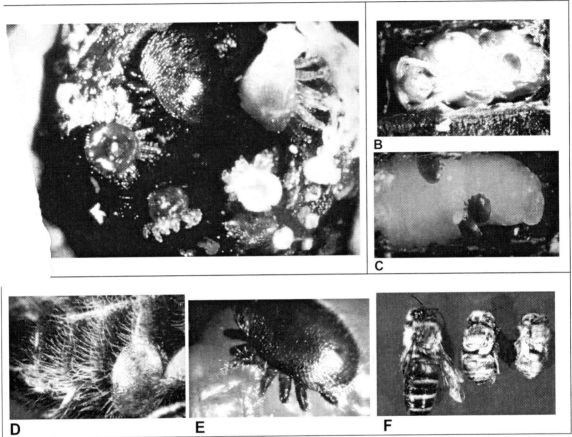

Figure 34-6. *Varroa* mites on honeybees. (**A**) Various life stages of the mites within a
honeybee cell (see also Figure 34-4). (**B, C**) Adult female mites feeding on young and older
pupae, and (**D**) on the abdomen of an adult worker; (**E**) a closer view of an adult female mite on
the ventral side of a pupa; damaged, (**F**) deformed workers that have eclosed despite being
hosts to *Varroa* mites.

Figure 34-7. Control
measures for parasitic
mites. (**Left**) MiteAway ®,
a formic acid fumigation
pad; (**right**) inserting
pesticide-impregnated strip
(e.g., CheckMite+® or
Apistan ®) into a hive.

There are a few other **ectoparasitic** arthropods on honeybees, but mostly they are regarded as inconsequential. On European honeybees the parasites *Acarapis dorsalis*, *Acarapis externis*, and *Acarapis vagans* are known, but they seem harmless. The bee louse, *Braula coeca*, superficially resembles the *Varroa* mite. It is really a kind of wingless fly (Diptera) that feeds at the mouthparts of adult honeybees, and seems to do no other harm than riding about on them. The maggots feed in the comb on pollen and debris.

Apis dorsata is host to a potentially very dangerous ectoparasitic mite. *Tropilaelaps clarae* is a fast moving, long-legged, but tiny mite. It transfers to *A. mellifera* colonies in places where both species of bee co-exist. It can be controlled by breaking the brood cycle, but its persistence in colonies of *A. dorsata* suggest that it can withstand some disruption in colony brood production.

Colony Collapse Disorder suddenly appeared in the U.S.A. as a major cause of dwindling of honeybee colony strengths at the end of the winter in 2007. The causes of this strange, yet devastating disorder have yet to be elucidated. The symptoms are decline and death of the affected colonies, but without corpses in or outside the hives. It is suspected that a combination of stresses, some resulting from diseases (mite and virus), others from management (especially from long distance migratory beekeeping), perhaps from pesticide use within and outside the hives, and from lack of diversity in diet.

...ง OF BEEKEEPING

There are several pests that get into honeybee hives and use hive products directly in their own life histories. They are **greater** and **lesser wax moths** (*Galleria mellonella* and *Achroia grisella*, respectively), the dried fruit or **bumblebee wax moth** (*Vitula emandsae*) and the **small hive beetle** (*Aethina tumida*). They have similar life cycles, even though the moths use different resources within the hive than does the beetle.

All three moths invade the hive as adult insects and lay their eggs in cracks and crevices in the hive and sometimes on the comb. The moths often invade stored equipment. Only the greater wax moth (Figure 35-1) can be considered a serious pest. The lesser and bumblebee wax moths overwinter as eggs, but that is not obligatory. The greater wax moth overwinters as a pupa, but again that is not obligatory. The small hive beetle originated in Africa and may not be able to tolerate freezing.

The eggs of the female moths can hatch over a long period of time. To a large extent, temperature is important in stimulating hatching. The larvae of the moths burrow into the comb. There they eat pollen, remains of larval honeybees, honeybee cocoon silk, and eclosed honeybees' faeces. . During their feeding, they spin layers of silk webbing over the surface of the comb (Figure 35-1). That is not only messy and difficult to remove for both the resident honeybees and the beekeeper, but it also stinks. The moth larvae pupate in tough, silken cocoons. After metamorphosis (Chapters IV – 12 & 14), they emerge as adults, mate, and the females disperse to find new oviposition places.   Greater wax moths have fascinating courtship songs combined with pheromonal calling.

Figure 35-1. (Left) The greater wax moth, *Galleria mellonella*, is about 13–19 mm long; (**Right**) the caterpillars produce webbing on comb. This example is mild in comparison to what happens in heavy infestations.

The lesser wax moth (Figure 35-2) is not as common in northerly areas, but can be a serious pest where the weather is warm. It is much smaller (10–13 mm) than the greater wax moth. The bumblebee wax moth is a sporadic and minor problem in parts of British Columbia, Alberta, and the Rocky Mountain States.

Figure 35-2. The lesser wax moth, *Achroia grisella*, on comb for scale. It is only about 10–13 mm long.

The small hive beetle (Figure 35-3) is a new and serious pest to North American beekeeping. It was first found in the USA in 1998 and presumably arrived from South Africa, where it is native and a minor nuisance to beekeepers. It includes honey in its diet. The adult females can lay up to 1,000 eggs each. The developing larvae invade the honeycomb, feeding and defecating there. Their activities cause the honey to dilute, ferment, and spoil. The odour of rotting oranges is associated with their feeding. Once they are ready to pupate, the larvae exit the hive, burrow into the ground, and pupate. The duration of the pupal stage is temperature dependent. Adult beetles have been recorded to live as long as 5 or 6 months.

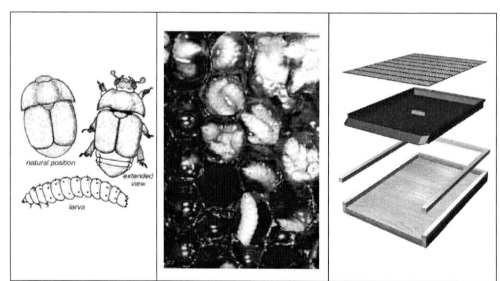

Figure 35-3. (**Left**) The small hive beetle, *Aethina tumida*. Adult beetles are only 5–6 mm long and dark brown. They normally assume a scale-like and protective posture. (**Centre**) Their grubs are highly destructive in honeycomb. (**Right**) Apart from chemical control, a West beetle trap, using a vegetable oil–filled pan, has been invented to catch the adults hiding in the hive and the larvae as they leave the hive to pupate.

Chapter 35
Pests of Beekeeping

The list of pests of beekeeping and beekeepers continues, but the above are the major culprits. Chapters IX – 33 & 34 deal with diseases, Chapter IX – 36 with predators, and Chapter IX – 37 with poisons. Those problems can also be thought of as pests, but they are separated because the bees are killed directly. Table 35-1 presents a list of pests and the means of dealing with them.

Table 35-1. Pests of Beekeeping, the Damage They Cause, and How to Deal with Them

Pest	Damage	Treatment
Greater Wax Moth *Galleria mellonella*	The most serious pest of stored equipment and can be damaging to active hives. They spin silken webs and smell bad.	Keep strong colonies; they can resist invasion by these moths. Cold treatment, freezing for 24 hours kills all stages. *Bacillus thuringiensis* (*Bt*) (Certan ®) works well but is no longer marketed. Chemical fumigation with paradichlorobenzene (moth balls or crystals) works well but beware that the chemical adsorbs into wax. Equipment so treated must be thoroughly aired.
Lesser Wax Moth *Achroia grisella*	Sometimes serious pests of stored equipment and can be damaging to active hives. They spin silken webs and smell bad.	
Bumblebee Wax Moth *Vitula emandsae*		
Indian Meal Moth *Plodia interpunctella*	These moths and some other pests (moths and beetles) can be problematic in pollen harvesting operations.	
Mediterranean Flour Moth *Agnagasta kuehniella*		
Death's Head Hawk Moth *Acherontia atropos*	African and European in range, it raids honeybee nests for honey. When caught, the adults squeak by using their proboscis and the caterpillars clash their mandibles.	Keep strong colonies; they can resist invasion by these moths.
Small Hive Beetle *Aethina tumida*	A recent invader to North American beekeeping. The larvae invade honey comb and ruin the honey. The larvae pupate in the ground.	The West beetle trap is reported to work well (see Figure 35-3). Chemical strips of the pesticide coumaphos (CheckMite + ®) can be used (as used for mite control; see Figure 34-7).
Other beetles	Various other beetles enter bee hives. Mostly they consume detritus. In Asia, a blister beetle may occasionally eat adult bees.	
Bee Louse *Braula coeca*	Generally harmless. The maggots burrow under cappings and eat pollen and debris.	
Termites *Isoptera*	These insects damage wooden hives in the tropics.	Hanging the hives out of reach of termites is commonly practiced in Africa. In Asia and tropical America, hive stands (for protection against ants) are used.
Mice and other small mammals	These animals may nest in hives in winter and so may destroy combs and frames.	Use of entrance reducers prevents their entry, and protects the hive from too much cold air.
Black Widow Spider *Latrodectes mactans*	May inhabit hives.	Caution when handling hives and equipment.

CHAPTER IX – 36

PREDATORS OF HONEYBEES

Predation is one of the fundamental interactions within any ecological community. A predator is an animal or plant that lives by preying on another animal. One of the main predators of honeybees is humankind, although human beings prefer to think of themselves in a more kindly light when it comes to interactions with honeybees. Table 36-1 presents a list of non-human predators of honeybees. Although most are inconsequential and worthy of notice mostly for the sake of appreciation of nature, a few can be problematic. The table makes note of how beekeepers can reduce their losses from them.

Table 36-1. Predators on Adult Honey Bees and Brood in the Hive

Predators	Damage	Notes on protection
INSECTS AS PREDATORS:		
Ants Argentine ants in southern North America Army ants in South America and Africa	Ants can be especially damaging in the tropics and subtropics. They attack and destroy entire colonies of bees.	Hives are placed on stands with their feet in cups of oil to act as ant barriers.
Hornets *Vespa* spp., especially *V. orientalis* and *V. mandarina* **Yellow jackets** *Vespa* spp.	Hornets forage in apiaries in Asia. These huge wasps may raid honeybee colonies in force; 20–30 hornets can kill 2,500 bees in a few hours and destroy, by raiding the brood, a colony of *A. cerana*. Yellow jackets attack bees, taking them as food for their own young. Sometimes they hunt at flowers.	*Apis cerana* guards "ball" the hornets and cook them to death with body heat (Chapter V – 19).
Other Wasps *Philanthus triangulum* European bee wolf, a kind of Digger wasp	*Philanthus* as a genus has a number of bee-hunting species. They patrol apiaries and rob colonies of brood and honey. In Europe it was estimated that one nesting 1.5 ha aggregation of the bee wolf collected 1.9 million dead honeybees.	
Other Insects	Occasionally other insects can be found inside beehives where they may damage comb and comb contents and even eat brood.	Some such insects may be beneficial in consuming dead bees.

Mantis	The predators can capture foraging bees at flowers.	
Spiders, especially **Crab spiders**	The predators can capture foraging bees at flowers.	
MAMMALS AS PREDATORS:		
Bears Black bear *Ursus americanus* Brown bear *Ursus arctos* Sun bear *Ursus malayanus*	These large mammals are destructive. In bee yards they smash the hives to get at the brood and honey inside. The black bear is the most troublesome in eastern North America. The smallest bear, and a skilled climber, hunts nests of *Apis dorsata* for food. This rare bear would probably forage in bee yards in tropical Asia.	The best prevention is an electric fence around the bee yard. Keep the grass around the fence short to prevent grounding by vegetation and allow the bear to be shocked. If bears are especially persistent, wildlife officers should be called in.
Honey badger or **Ratel** *Mellivora capensis*	In parts of Africa these animals raid and destroy hives. They are reported to have a relationship with the honeyguide, a bird that locates wild honeybee nest and directs the ratel to the site (see below). The ratel opens and destroys the hive; the bird gets the leftovers.	Hanging hives from trees or stands prevents these animals from destroying them.
Skunks *Mephitis* spp. but mostly *M. mephitis.* *M. macroura*, the hooded skunk, occurs in the southwest USA and Mexico	These nocturnal animals visit hives and scratch at the entrances. The bees that come out are eaten. The skunks may become regular, nightly visitors to a bee yard.	A hive entrance board of sharp nails can discourage the skunk. A raised entrance of metal grillwork (hardware cloth) allows the bees to get beneath the skunk and sting its belly. Some beekeepers poison skunks with bait in the bee yard.
BIRDS AS PREDATORS:		
Kingbirds (*Tyrannus* spp.) and **Martins** (*Progne* spp.)	Insects are a natural part of many birds' diets. These birds can catch bees in flight.	In the past, shooting was suggested. Nowadays, with such birds having declining population, conservation is recommended. They are never major problems.

Chapter 36
Predators of Honeybees

Woodpeckers (*Piciformes*)	Woodpeckers may attack hives or domiciles of other bees when hard frosts restrict their foraging. They may damage hives and domiciles.	
Bee-eaters (Meropidae)	These colourful birds are specialists at eating stinging Hymenoptera. They are tropical and subtropical. They may become familiar with the location of bee yards and become serious pests.	These birds catch their prey on the wing. Stringing a grid of fishing line about 2–3 m above the bee yard may prevent them from hawking bees close to the hives.
Honeyguides (*Indicator* spp.)	These birds live south of the Sahara in Africa. After the ratel, or other honeybee nest predator, has raided, the birds consume any leftover brood, honey, wax, and combs (see above).	Honeyguides get their name from their behaviour in attracting the attention of mammals to the location of bees' nests.
Honey buzzard (*Henicopernis longicauda*)	This bird of prey attacks the colonies of *Apis dorsata* by flying at them to disturb the bees. The bees follow the buzzard away from the nest. The buzzard returns to feed on brood. These birds have been seen removing the lids from hives of *A. cerana* and removing the frames in their talons and eating.	
Swifts (Apodidae) and **Flycatchers** (Tyrannidae; includes kingbirds (above))	These birds can catch bees in flight.	
TOADS AND LIZARDS AS PREDATORS:	Occasionally these animals can ambush adult bees at the hive entrance. Some small lizards take up residence in the protected confines of a hive.	Remove to another location if really necessary. Their depredations are probably inconsequential.

CHAPTER IX – 37

POISONS

The major cause of poisonings of honeybees is through application of agricultural—and, to a lesser extent, forestry—pesticides. Pollution can be a problem under special conditions. There are even natural sources, in nectar and pollen, of chemicals that poison honeybees.

The pesticide problem and bee kills have been around since chemical pesticides, such as arsenicals, were first introduced into agriculture in the 1880s. Newer, synthetic organic chemicals came into widespread use after World War II. Organochlorides, such as DDT, have become notorious for their environmental effects even though they are still useful in some circumstances. Dieldrin was the first to cause major kills of honeybee colonies in the 1950s. The organophosphorous insecticides were developed a little later, but soon were reported as dangerous to beekeeping. Parathion and malathion were reported killing honeybees in the late 1940s. The carbamate insecticides are also poisonous to bees, and carbaryl has a long black history in beekeeping circles. Pyrethroids (they mostly end in "thrin") may be less damaging to beekeeping, but are, nevertheless, broad spectrum insecticides (i.e., they kill lots of things). Most recently, a variety of new insecticide types have been developed, such as the chloronicotines, spinosyn, and avermectin. Table 37-1 notes the classes of chemicals, their mode of action, and the symptoms they cause in poisoning honeybees.

Herbicides, by and large, are not directly poisonous to bees. They kill plants. Some may become incorporated into nectar, but instances of poisonings through that pathway are somewhat conjectural. The adverse effects of herbicides are in the reduction of the amount of bee forage in sprayed locations. Fungicides are mostly even less of a problem to beekeeping.

Pollution and smog affect life in general, but their effects will likely damage people and plants before they damage honeybees. There are instances of pollutants in the soil and air becoming incorporated into plant tissue and secretions, such as nectar. After the nuclear power plant in Chernobyl blew up, radioactive materials were found in trace amounts in nectar and honey. Honeybees are good samplers of the air through which they fly. Some pollutants adsorb onto their cuticular wax (Chapters I – 1 & IV – 12) and can be detected. Similarly, bees' wax takes up some materials from the environment (Chapter X – 39).

The chemicals that are used by beekeepers (Chapters IX – 33 to 35) need to be applied sparingly. Some are toxic to bees; some adsorb into bees' wax; some contaminate honey.

Chapter 37
Poisons

Table 37-1. The Main Classes of Insecticides, Their General Mode of Action (How They Work), and Their Effects on Honeybees

CHEMICAL GROUP	MODE OF ACTION	EFFECTS ON HONEYBEES
Chlorinated hydrocarbons DDT, dieldrin, aldrin, chlordane, heptachlor, lindane are all highly toxic to bees. Use of these chemicals is now curtailed.	Interfere with nervous transmission (Chapter III – 9).	Bees tremble and have erratic behaviour; they often drag their hind legs and hold their wings hooked together and out to the side. Death tends to be slow, and many poisoned bees fly and die away from the hives.
Organophosphates Related to nerve gas	These chemicals inhibit activity of cholinesterase, an enzyme involved in breaking down acetylcholine in synaptic nervous transmission (Chapter III – 9), so death is by nervous exhaustion.	Bees regurgitate and may become wet with honey or nectar. They appear erratic and disoriented. They often have distended abdomens, hold their wings hooked together and out to the side, and extend their tongues. Death is often within the hive.
Carbamates		Causes erratic behaviour, stupefaction and paralysis. Carbaryl (Sevin) is especially noted to cause poisoned bees to become unable to fly and to crawl about on and in the hive as they die. Many bees die in the hive. Chronic sublethal poisoning disrupts colony activity and brood cycle.
Pyrethroids	These chemicals interfere with the sodium balance of the nerve cells, which is crucial to transmission of nerve impulses (Chapter III – 9). Rapidly repetitive nerve firing leads to nervous exhaustion and death.	Bees regurgitate and may become wet with honey or nectar. They become paralysed. Many die in the field.
Chloronicotines Imidocloprid	These chemicals bind to the acetylcholine receptor in nervous synapses, thereby disrupting transmission.	In very small, sublethal, amounts, foraging behaviour may be disrupted. The problem is not fully understood. Highly toxic to bees.
Avermectin	The chemicals interfere with neural and neuromuscular nervous impulse transmission.	Highly toxic to bees.
Spinosyns Complex molecule derived from *Saccharopolyspora spinosa*, a soil fungus.	These chemicals act like acetylcholine and cause continual stimulation across neural synapses.	In very small, sublethal, amounts, foraging behaviour may be disrupted. The problem is not fully understood. Highly toxic to bees.

For beekeepers to avoid pesticide poisonings, and for pesticide users to avoid the wrath of beekeepers, co-operation is required. Pesticide labels carry warnings about toxicity and hazards to honeybees. Applicators should adhere to the instructions on the labels: those instructions are the law. Common sense goes a long way to preventing problems. Honeybees, in fact most bees, forage from mid-morning to late afternoon, so applications made early or late in the day reduce the risk of poisoning. Beekeepers can assist by making the locations of the bee yards known to pesticide applicators and neighbouring growers. They can leave contact information with them, and at the bee yards. That way, applicators and growers can warn beekeepers of forthcoming applications. The best protection for beekeepers is to try to have bee yards that are distant from where pesticide sprays are likely. Beekeepers can take action by moving their bees away, temporarily closing the colonies, or covering the hives with wet burlap to confine and protect the bees while keeping them cool. Although those precautions seem sensible, they are often impractical.

When pesticide poisonings of honeybees are suspected, the beekeeper should obtain corpses of the bees that are as fresh as possible and store them by freezing at as low a temperature as possible. Then the corpses should be delivered to a pesticide residue analytical laboratory as quickly as possible, with information on when and where the sample was collected and what sort of pesticide is suspected. Remember that pesticides break down in nature, and in rotting bees, and the analytical chemist uses different methods to detect different chemical residues.

The science of toxicology as applied to honeybees is well developed. It involves treating honeybees in various ways with pesticides and determining how much causes death. The poisons are administered topically (on the body) by spray or on surfaces that the bees must walk on; orally (as food); or by fumigation (as breathed).

Natural poisons occur in some nectars and pollens. The California buckeye (*Aesculus californica*), locoweeds (*Astragalus* and *Oxytropis* spp.) and death camus (*Zygadenus* spp.) secrete nectars that can kill bees. Even almond (*Amygdalus communis*), known to be excellent honeybee forage and very dependent on honeybees for pollination, has nectar and pollen rich in amygdalin, known to be toxic to honeybees. Although poisonous nectar and poisonous honey is known, it seems that the poisons are reported to affect people more than they affect the bees (Chapter X – 38).

Section X

Beekeepers' Rewards
Honey, Bees' Wax, Pollen, Royal Jelly, Propolis, Venom, More Bees & Pollination

Beekeepers work hard. They invest money, blood, sweat, and tears in their operations. They expect returns: money and satisfaction.

Most beekeeping operations, as well as hobby beekeepers, enjoy **honey**. It is a most interesting food, with many values beyond just being a sweetener. It has medicinal value, and is used in the baking trade for its additional qualities in improving texture and shelf life of baked products. It is processed in bulk and in small lots. It may granulate and is often solidified for sale. It is traded worldwide. There are standards for processing, sugar content, assessing floral sources, and contaminants. Human reverence for honey is culturally almost universal, and very ancient.

Bees' wax also enjoys the respect of ancient traditions. Its values in illumination (candles), cosmetics, medicine, pharmaceuticals, and as a waterproofing and lubricant have long histories. The complex chemistry of bees' wax is what gives it its properties, and it has not been artificially made.

The other products of the hive, **pollen**, **royal jelly**, **propolis** (bee gum), and **venom** have value in more recent developments in **apitherapy** (use of bees and bee products in treating ailments). Because hive products are so revered and considered pure, and because beekeepers can recoup investments by sales of those commodities, markets in "health food" and natural medicines take up supplies of pollen, royal jelly, and propolis. The collection and sale of **venom** is mostly for medical use in bee-sting therapy for desensitization of allergies, and treatments of mostly inflammatory ailments.

The production of more bees is vital to the beekeeping industry. **Queen rearing** is a specialized occupation requiring an intimate understanding of the life history of honeybees. Queen breeders are involved in selection and maintenance of quality lines of honeybees that sustain the industry. There is even a small market for bee brood as food for people and other animals.

Without doubt, the greatest value that honeybees contribute to food production and security is through their activities as pollinators of crops. Beekeepers who provide **pollination services** to growers must assure the growers that their colonies are as strong and active as can be expected for a given crop at a particular time of year. Managing honeybees for pollination requires additional skills, and knowledge of floral biology. With pollination being increasingly recognized as a crucial process in agricultural intensification and production, and wild pollinators becoming scarce in many places, beekeepers are finding themselves much in demand. Moreover, there is global and international concern for the demise of pollinators, both wild and managed for their services to plant reproduction in nature and in agriculture.

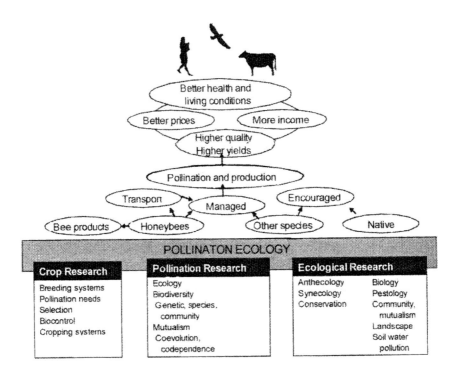

The relationships between pollination, human and natural affairs, and research and development.

CHAPTER X – 38

HONEY

Honey is the most well recognized and appreciated of hive products. It is a sweet viscous liquid prepared by bees, mostly honeybees, but also stingless bees and even bumblebees, from various sugar sources, mostly plant nectaries, and stored by them for food.

Honeybees add enzymes from their salivary glands at the time of nectar collection (Chapters II – 7 & IV – 13). Water is evaporated from the nectar, and the final amount of water in honey varies depending on climate, temperature and humidity. The amount of water in honey is important in quality control and for assessing likelihood of fermentation (see below).

Honey is highly variable. Its taste, colour, aroma, and texture depend not only on the region where it originates, but also on the source of the sugary forage the bees use to produce it. Floral nectar is the most important source of honey. There are also extra floral nectaries on many plants, and bees are quick to discover and use them. Similarly, bees find honey dew of insects and plant origin. Aphids and scale insects excrete large quantities of excess sugary liquid while processing dilute plant sap for their own nutrients. A number of plant pathogens also produce sugary exudates that bees exploit. Occasionally, bees take the juices of fruits, to the extent that they are sometimes minor pests.

The **honey types** marketed by beekeepers can reflect **floral source** (e.g., clover honey, blueberry honey, linden honey, orange blossom honey, etc.) or can be a **seasonal blend** indicating the mix of floral sources that are abundant in the spring, in the early, mid-, and late summer, and in the fall. The type can also reflect the way in which the honey is prepared or obtained. It is then named after the method of preparation and the honey's properties. **Liquid honey** is just that. It may crystallize after it has been sold, but that does not detract from the nutritional qualities of the honey. **Crystallization** of honey can be highly controlled so that it is sold as solid, **creamed honey** that can be spread easily and without dripping from bread, scones, etc.

Honey may be sold in the comb. **Section comb honey** is specially produced in wooden frames inserted into the bee hive. **Cut comb** is cut from frames that were originally equipped with thin foundation without wire supports. The cut pieces are sold in plastic or aluminum trays. **Chunk comb honey** is attractive, but not often seen. Chunks of cut comb are placed in jars, and then liquid honey poured in to fill.

Most honey on the market, though, is sold as **bulk liquid honey** for processing by **honey packers**. That is how it reaches the shelves of supermarkets and makes its way to the food industry for baking, confectionary, and brewing.

There are internationally recognized **industrial standards** for honey quality and grading as follows:

Moisture content: <18.6% Grade A (described as Fancy)
                                20% Grade C   (described as Standard)

Colour by the Pfund Scale and based on light transmission:
                  Water White
                  Extra White – Clover
                  White
                  Extra Light Amber
                  Light Amber
                  Amber
                  Dark Amber – Buckwheat

There are other standards that must be met. Those are set to reflect the quality of the honey, the care with which it has been handled, its floral origins, and its age; and to check that the honey has not been adulterated with other syrups (notoriously corn syrup), and that it is free of residues such as of pesticides, drugs used by beekeepers, or metals from processing equipment.

The **moisture content** of honey is usually measured by its **refractive index**. As lights passes through some solutions it is bent, as in a glass prism or lens. The extent to which light is bent, or refracted, depends on the amount of sugar present. The units are **brix**, or **% sugars**. High quality honeys are clear, although the darker the colour the less clear they are. The **clarity** of honey can be improved if it is allowed to stand and air bubbles and small particles of wax are allowed to float to the surface. **Turbidity** is the converse of clarity, and in honey quality assessment and judging, low turbidity indicates careful handling and processing.

With so much sugar dissolved in so little water, honey is dense: 1 litre weighs 1.42 kg. Also, sugar is a supersaturated solution of sugars, so it may granulate (see below), but has a tendency to be hygroscopic, that is, it absorbs moisture from the air if the humidity is high (Figure 38-1). Certainly in humid places, the bees are unable to evaporate moisture from honey to produce even standard grade honey. The result is that honeys from humid areas, particularly of the tropics, have a tendency to ferment rapidly, and spoil.

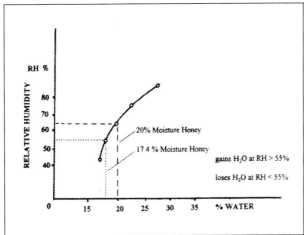

Figure 38-1.
Hygroscopicity of honey with various water contents under varying conditions of atmospheric relative humidity. Fancy grade honey absorbs atmospheric moisture at relative humidities above about 55%.

Honey is viscous, that is, it flows slowly. As the concentration of sugars increases, the viscosity of honey increases. Also, as the temperature of honey drops, honey becomes more viscous. The units for viscosity are "poise"; water has a viscosity of about 0.01 poise (P). The viscosity of honey also depends on the floral source, and the sugar constituents. Sage honey is more viscous (115 P) than is clover honey (94 P), and clover honey is about 70 times more viscous when its moisture content is about 15% (138 P) versus 20% (20 P). Sweet clover honey has a viscosity of about 68 P at about 30°C, but 190 P at room temperature. These characteristics are important in honey packing because efficiency requires that pumping honey be inexpensive and that containers be filled quickly.

Honey should not be kept at high temperatures (above 50°C) for long because it tends to denature. Hydroxmethylfurfuraldehyde (HMF) is the breakdown product of sugars and can be measured for quality assurance. Fresh honey normally has about 10 mg HMF/kg of honey; heated and stored honey may have 30–40 mg HMF/kg. And honey adulterated with invert sugars (sugars that have been heated in the presence of acid) may have as much as 150 mg HMF/kg of "honey". HMF is not harmful, and many syrups contain much more than is usually found in honey.

The chemical composition of honey varies, mostly as a consequence of the origins of the nectars from which the bees make it (Table 38-1). It is the minor constituents that give honeys their characteristic odours and flavours. The aromatic compounds are mostly derived from flowers. Hence, for example, orange blossom honey has the characteristic scent of citrus.

Table 38-1. Chemical Composition of Average Honey

Major constituents (about 99%)	Percent/Notes
Water	17.2
Glucose (Dextrose)	31.3
Fructose (Levulose)	38.2
Sucrose	1.3
Disaccharides (Maltose)	7.3
Dextrins and gums	1.5
Minor constituents (about 1%)	
organic acids (formic, acetic, malic, citric, succinic, lactic, tartaric, oxalic, carbonic, pyroglutamic, gluconic)	0.57 Honey is acidic: **pH 3.6 to 4.2**
minerals (potassium, chlorine, sulphur, sodium, calcium, phosphorus, magnesium, silicon, iron, copper)	0.17
nitrogen (in amino acids and proteins), (albuminoids and colloidal proteins, amino acids)	0.04
plant pigments (carotin, xanthophyll, tannin, anthocyanin, chlorophyll, decomposition products)	trace
bees' wax and foreign particles (yeasts, molds and spores, parts of bees, etc.)	variable
Vitamins	mostly B from pollen
Enzymes: Invertase Glucose oxidase Diastase (amylase) Catalase Inulase	- converts sucrose to glucose and fructose - oxidizes glucose to gluconic acid - converts starch to maltose - decomposes hydrogen peroxide - converts inulin to fructose
Aromatic substances (terpenes, aldehydes, esters, higher alcohols)	Hydroxymethylfurfuraldehyde is used to test for denaturing by heat and age

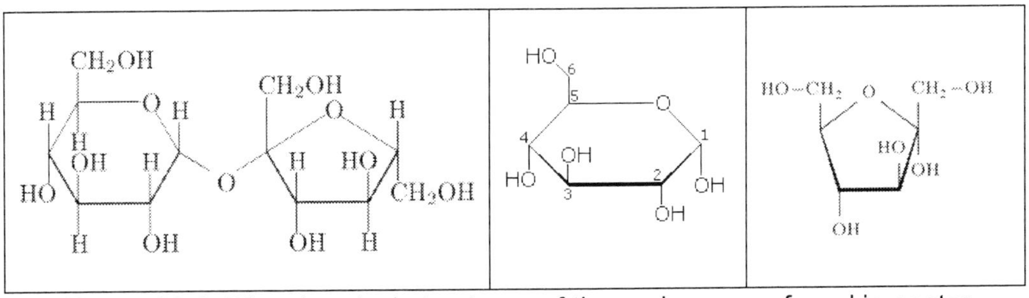

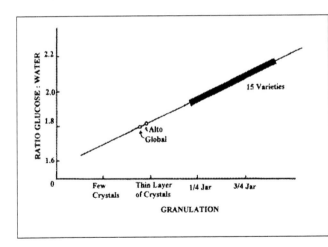

Figure 38-2. The chemical structures of the main sugars found in nectar and in honey: sucrose (or saccharose); glucose and fructose (alpha-D-fructose)

The various constituents of honey give it its properties. Sucrose, although abundant in many floral nectars, is broken down into the two most abundant sugars in honey—glucose (or dextrose) and fructose (or levulose)—by invertase that is secreted by the bee's salivary glands (Figure 38-2). **Glucose** (dextrose) is the sugar that crystallizes to form granulated and creamed honey. If honey has a **ratio of glucose to water** of more than about 2, the honey tends to **granulate** (Figure 38-3). Glucose is not as sweet as fructose, but is an important energy source for bees and for people. It is chemically stable.

Figure 38-3. The expected amount of granulation in jars of honey derived from nectar of canola flowers. Based on the amount of glucose in the nectar, most types of canola honey would readily granulate. That can be a problem for honey storage.

Under examination by a polarimeter, glucose rotates polarized light to the right. **Fructose** (levulose) is the sugar that gives honey much of its characteristics and flavour. It is very sweet, hygroscopic, does not granulate, and when examined by polarimetry, rotates polarized light to the left. The **organic acids** are part of the reason that honey is acidic. The pH of honey ranges from 3.6 to 4.2 (vinegar, in comparison, is 3.1), so honey would taste sour if not for its very high sugar content.

The **mineral content** of honey reflects the common elements of the soils and nectars of the region. Generally, the mineral content is higher in darker honeys.

The minor presence of **proteinaceous materials, pigments, bees' wax,** and so on provide distinctive texture, taste, aroma, and colour to honeys (see Table 38-1). Even bitter and poisonous honeys exist, especially as derived from *Rhododendron* flowers. Sometimes, the presence of proteinaceous materials in honey can impart characteristic properties. **Thixotropy** is the tendency of a liquid to form a gel. In honey, the presence of colloidal substances causes gelling. Heather honey is thixotropic. A similarly caused characteristic is **stringiness** in honey: when it is pulled from the container it forms hair-like threads. Neither of those properties is harmful and both can be reversed by stirring the honey.

Pollen grains are present in almost all honeys. The average number for Canadian honeys is approximately 4,500,000 per kg. These probably contribute to the flavour and aroma of honeys. Their presence is used to check on the validity of claims of "floral sources". Honey tested by melissopalynology (the study of honey) is first diluted, then centrifuged to obtain a pellet of the particulates. That pellet is then examined under the microscope and the pollen grains found are identified (Figure 38-4).

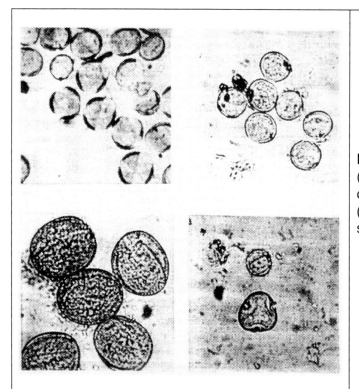

Figure 38-4. Pollen grains in honey: (**Top left**) Canola, (**Top right**) White clover, (**Bottom left**) Buckwheat, (**Bottom right**) Golden rod (not to same scales).

Honey is not a benign environment for microorganisms. The high concentration of sugars prevents growth because the sugars draw water, by **osmosis**, from the microbial cells. Nevertheless, all honey contains **osmophilic** (sugar-tolerant) yeasts which can multiply, with concomitant fermentation, if the water content is high and the temperature favourable. Those yeasts feed on the sugars and break them down to form alcohol. As a general rule, honey's tendency to ferment is an interaction between the amount of water and the number of yeast cells present (Table 38-2).

Table 38-2. The Moisture Content of Honey, Number of Yeast Cells Present, and Likelihood of Fermentation

Moisture Content	Yeast Cell Count	Likelihood of Fermentation
> 17.1%	Safe regardless of yeast cell count	
17.1–18.0%	yeast < 1000/g	safe
	yeast > 1000/g	ferments
> 20%	any yeast at all	fermentation always

Honey is a high energy food. It promotes resistance to fatigue. It is reported to aid in digestion and help protect from infection, and it has been used successfully in treating wounds in people and animals. Its antimicrobial properties are well documented, but not understood. It has been used to alleviate eye irritations because of its lubricative and medicinal activity. Honey is widely used in medicine, especially Ayurvedic treatments in which it is used as a carrier for herbal remedies. It is used widely in cooking. Honey improves texture, moisture retention, and shelf life of baked goods. Honey is often included in confectionery recipes, including breakfast cereals, snack bars, spreads, marmalades, and jams. It can be mixed with dairy products, such as milk, yogurt, and ice cream. It is used in various non-alcoholic drinks, and may be fermented to make mead (honey wine) and honey beers. It is even used in the tobacco industry to improve the aroma and moisture of pipe and cigar tobacco.

Outside direct human affairs, it is used in feeding insects raised for experimental purposes and for biological control.

Honey is also a revered food associated with plenty and with celebrations (Chapter VIII – 31).

CHAPTER X – 39

BEES' WAX

Wax describes a wide variety of materials that are solid and slippery to touch. They occur naturally on plants and animals, and are made from petroleum. Natural waxes are not pure compounds. They are a mixture of long chain fatty acids with other constituents that all vary according to their origins. **Bees' wax** is no different. It is secreted by worker honeybees about 9–17 days old, and used by the colony to construct combs. It comes from four pairs of wax glands on the underside of the abdomen (Chapters I – 4 & II – 7).

The production of bees' wax by the honeybee is not well understood, but bees' wax is costly for the bees to make. Honeybees consume approximately **6.5–9 kg** honey to make **1 kg wax**. It is synthesized from the sugars of honey eaten by the bees and requires protein for the production of enzymes to catalyze the processes. It has been suggested that probably esters are synthesized in the **fat body** (Chapter II – 8), and the hydroxy acids and paraffins are produced by *oenocytes* which are also embedded in the fat body.

Wax-making bees are usually quite young (Chapter IV – 15). They fill themselves with honey and hang in clusters on the comb. In that position they synthesize the wax within their bodies and pass it out through the wax glands into the abdominal wax pockets as eight small, shiny white flakes. Those are gathered up, and chewed by the comb-building, or cell-capping, bees as it is used in construction.

Figure 39-1. (Left) Liturgical candles from St. Brigid's, Kilburnie, Scotland, and (**right**) "dripped" candles from Cheeky Bee.

People have used bees' wax for many purposes for millennia. Perhaps its most commonly appreciated use is for candle making (Figure 39-1). It has been used since ancient times for cosmetics, medicinal ointments and creams, embalming, candlelight, writing tablets, lubrication, sealing, adhesives, varnishes, polishes, art as part of encaustic painting and other methods of applying colours to surfaces in painting, sculpture (e.g., in wax museums), etching, and metal casting by the "lost-wax" method. More modern uses have been in dentistry for making pallet and gum molds for denture production, plumbing seals, armament (bullet) waterproofing, crayons, chewing gum, horticultural grafting paste, sport waxes (cross-country skiing), electronics for its dialectric and waterproofing properties, and recycling within the beekeeping industry by making foundation (Figure 39-2; Chapters VIII – 31 & 32). For the last mentioned use, care must be taken in preparation to avoid transmission of spores of American Foul Brood (Chapter IX – 33).

Figure 39-2. A foundation press and cutting machine.

Bees' wax has many useful **physical properties**. It does not mix with, or absorb, water, and can form a protective layer to make a surface impermeable to water. It becomes plastic at about 32°C, so can be formed by the bees' chewing on it at usual temperatures in the hive. That property also makes it a convenient "chewing gum", and it can be worked, shaped and carved by human hands. Its great plasticity and strength over a wide range of temperatures makes bees' wax extremely useful. It has literally hundreds of uses.

Its melting point ranges from 61–66°C and it solidifies between 60–63°C. Thus, when just molten, it can be handled without much risk of burning. Thus it can be used in liquid form for molding, making foundation, and sealing.

It is soluble in ether, benzene, and chloroform, a property that allows some industrial applications. It burns slowly and steadily when made into candles.

With its relative density (specific gravity) of about 0.96 at room temperature, it is less dense than water. A simple way of melting and collecting bees' wax is in hot water. The wax floats to the surface. When the hot water and wax cools, the floating wax solidifies and can be removed as a cake.

Chemically, bees' wax is mostly a mixture of normal paraffins, each consisting of a series of long-carbon-chain compounds with acid or alcohol (or both) moieties. Over 284 compounds have been identified from bees' wax. Many of those are volatile, and some contribute to the characteristic, honey-like odour. They are lost in bleaching. Table 39-1 presents the more important constituents.

Table 39-1. The Most Prevalent Compounds in Bees' Wax

General Type of Chemical*	Comments on Chemical Diversity	%
Hydrocarbons (paraffins)	C_{21}–C_{33} (mostly with odd numbers of carbon atoms in the chain: some 10 major and 66 minor components have been identified)	14
Monoesters	C_{24}–C_{36} (even numbers of carbon atoms in the chain: some 10 major and 10 minor components have been identified)	35
Diesters		14
Triesters		3
Hydroxy monoesters		4
Hydroxy polyesters		8
Acid esters		1
Acid polyesters		2
Free acids	Mostly hexadecanoic (palmitic) acid C_{16}	12
Free alcohols		1
Others (unidentified)		6

*Details for paraffins and monoesters represent the general complexity of the array of chemicals.

Bees' wax is collected in various ways. The highest quality originates from the cappings of honey comb collected during extraction (SECTION VIII). Processing about 1 tonne of honey yields about 10 kg of wax. Wax can be salvaged from old and damaged combs, but it tends to be dark and requires filtering and bleaching before it can be re-used. In parts of Africa, bees' wax is harvested from honey comb as it is crushed to squeeze out the honey (see also Chapter VIII – 31). The wax is made into cakes (Figure 39-3) by the use of hot water (as described above). It should not be heated above 85°C because otherwise it discolours. The containers used for harvesting bees' wax should not be iron, zinc, copper, or brass because the wax and the acidic honey reacts with those metals, causing discolouration of the wax.

Figure 39-3.
Cakes of bees' wax (about 1 kg each) in storage.

Bees' wax is a natural absorbent for many pesticides. For that reason, quality control of residues is crucial for wax destined for pharmaceutical and cosmetic uses. Propolis mixed in with bees' wax also lowers the quality. For those reasons, the highest quality bees' wax does not originate from the industrialized and highly agricultural world. Bees' wax cakes are durable, and easily stored and transported so can be an important source of revenue for beekeepers and honey gatherers in remote places.

The waxes of Asiatic species of honeybee, "ghedda waxes", are described as being softer and more plastic. Nevertheless, the waxes from European and tropical African *Apis mellifera* and from *A. cerana, A. dorsata* and *A. florea* are all similar in composition. Bees' wax in Asia is used extensively in batik art and printing. Cloth or other surfaces are treated with wax, which is removed where pigment, or dye, is wanted. Intricate and beautiful works of art, clothing, tablecloths, and so on are traditionally made in India and Southeast Asia by this method. The wax of bumblebees is softer than that of honeybees. Bumblebees mix pollen with their wax to give it greater strength in nest construction.

CHAPTER X – 40

POLLEN

Individual pollen grains are the **male gametophytes** (haploid stage N) of the flowering plant (Figure 38-4). They are tiny grains, from 6 to 200 microns in diameter, that range from smooth to highly ornamented with textural patterns and spines. Some are dry, others are oily. They are produced in the anthers of flowers and their role is to be transferred to the stigma of another or the same flower. There, if they are on a compatible stigma (i.e., of the same species of plant or of a compatible genotype (Chapter X – 45) they germinate. Germinating pollen grains produce a pollen tube that grows through the stigma, into the style, and eventually into the ovary of the plant (Chapter X – 45). While the pollen tube is growing, one of the nuclei inside divides to form two sperm nuclei. When those are released into the ovary and find an ovule, double fertilization occurs. One nucleus joins with the egg nucleus to form a zygote and then embryo; the other combines with other cells of the ovule to form the endosperm. The endosperm becomes a nutrient-storing tissue in the seed and, when the seed germinates, it feeds the growing embryo.

Many pollinators, especially bees, rely on pollen as a source of food for themselves or their offspring (Chapter IV – 13). During the co-evolution of flowering plants and insects, plants have adapted to providing pollen to their pollinators. The pollinators transport the pollen during the process of pollination, which must occur before fertilization (above).

Bees are adept at harvesting pollen (Chapter I – 3; Figure 40-1). Honeybees have specialized corbiculae into which they pack pollen for transport back to their hives. If the pollen is dry, bees moisten it with sugar from honey or nectar to pack it into **pollen loads** on their hind legs. Pollen so packed is removed from circulation in pollination and is mostly inviable. So-called "bee pollen" has become an important commodity in the natural foods market and for apitherapy.

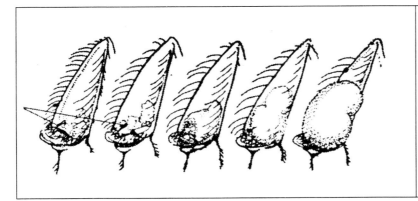

Figure 40-1. The accumulation of pollen in the corbicula (pollen basket) of a honeybee or bumblebee (see also Chapter I – 3).

Pollen is highly variable in composition. It is chemically complex, and most plants' pollen is highly nutritious to bees and other insects. It contains, on average, 25% protein, 10% free amino acids, 25% carbohydrates (including starch), and the remaining percentage made up of varying amounts of lipids, enzymes, co-enzymes, pigments, vitamins, sterols (Chapter IV – 14), and minerals. The wide variation in the chemical constituents of pollen reflect the diversity of plants from which it originates.

Pollen is collected from honeybee colonies by the use of **pollen traps** placed on the bottom board and under the rest of the hive, or in the entrance to the hive. There are a number of designs, but they all work on the same principle. As the returning pollen forager enters the hive, it must crawl through a double coarse screen barrier. The bee can fit through, but not with its pollen baskets loaded and out to the sides like fat saddle-bags. The pollen loads are knocked off as the bee crawls up and into the hive. The screens used have 5 wires per inch (2 per cm) and are 4.7 mm apart. One of the first, and most successful, pollen traps was developed at the Ontario Agricultural College and is referred to as the **OAC pollen trap** (Figure 40-2). It fits atop the bottom board. It has a tray of fine mesh beneath the two pollen trapping screens. The mesh allows for ventilation and drying of the trapped pollen loads (pollen pellets). The mesh tray opens as a drawer from the rear of the hive to allow easy access and emptying. Another feature of the OAC trap is drone exit ways, through which workers can also leave the hive. The exit holes are small so that few returning foragers find them to re-enter the hive.

Figure 40-2. An OAC pollen trap viewed from the back. The double trapping screens can be seen as the top; the harvesting tray is partially withdrawn.

Trapped pollen remains as pellets in the form of the loads brought back by the bees (Figure 40-3). It comes in a variety of colours, predominantly yellow and orange.

Figure 40-3. A small sample of pollen pellets (loads) collected in a pollen trap.

The amount of pollen collected by a colony depends on the colony's immediate requirements (Chapters IV – 13, 15 & V – 18). Using the pollen traps increases the number of bees collecting pollen but reduces nectar collection. Given that it is estimated that a colony of honeybees requires 20–30 kg of pollen for its own needs each year, one may understand that pollen trapping can stress colonies nutritionally. Nevertheless, annual harvests of pollen taken from honeybee hives range from 25–55 kg, with up to 2 kg being collected in a single day from populous colonies.

Once pollen is collected, it must be dried thoroughly. It can mold and ferment if damp, and if frozen damp becomes caked. For human consumption, beekeepers must be aware of the dangers of pesticide contamination from agricultural sprays.

The dried pollen is used for human consumption, either as capsules or in cookery. It is reported to improve stamina, appetite, vitality, sexual prowess, and haemoglobin levels in blood. It has been fed to race horses, too. Reported medicinal properties are in treatment of colds, anaemia, acne, high blood pressure, nervous and glandular disorders, ulcers, male sterility, and even cancer. Pollen consumption has been advocated to aid in the reduction of pollen allergies. Pollen extracts have been used in some cosmetics for skin rejuvenation.

Although pollen clearly can be regarded as nutritious, the claims of its other benefits are not substantiated scientifically.

Other uses for bee-collected pollen are in the honeybeekeeping industry. It can be used as supplemental food for colonies in spring or winter. Because of the risk of transmission of bee diseases with bee-collected pollen, it is irradiated during processing so that it is safe for sale and use. Honeybee-collected pollen is also used in bumblebee husbandry (Chapter XI – 48). It is used in the bumblebee-rearing houses to allow the queens to initiate their colonies and feed their brood.

CHAPTER X – 41

ROYAL JELLY

Royal jelly is fed to all larval honeybees, and to the adult queen (Chapter IV – 13). It is mostly the secretion of the hypopharyngeal glands, and a small amount from the mandibular glands in the heads of young (5–15 days old) workers (Chapter IV – 15). It may contain some regurgitated honey from the nurse bees' crops, mainly sugars. Thus, when analyzed, it has been found to be quite variable in its constituents. Interest in production of royal jelly seems to have started in the 1950s with claims about its value in apitherapy and properties as a panacea. There is a thriving market for royal jelly. Most production comes from the Far East.

It is a creamy, milky white, acidic (pH 4.1–4.8), slightly pungent thick liquid (Figure 41-1) with a bitter to sour flavour. Its chemical composition (Table 41-1) shows that it is a rich food, highly proteinaceous, with lipids, sugars, vitamins, sterols, and minerals.

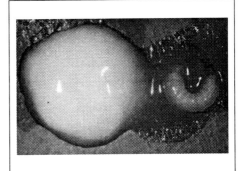 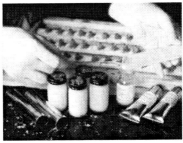

Figure 41-1. (Left) Royal jelly, with larval honeybee off to the right. **(Right)** Collecting royal jelly for scientific research. Note the frame of queen cups in the background.

Production of royal jelly requires beekeepers to maintain colonies in a high state to nutritional health, and to manage the colonies as if they were raising queens. Royal jelly is secreted most actively into queen cells. Beekeepers producing royal jelly use artificially made queen cups placed along bars on specially made frames (Figures 41-1 (right) & 41-2). Well-managed hives can produce about 500 g of jelly over about 6 months. That represents production from about 1,100 queen cups. Its production requires fastidious management, precise timing, and cleanliness. Royal jelly production is labour intensive, in the bee yard and in the harvesting facilities.

The queen lays eggs into those cups, and they are harvested after the larvae are 4 days old, when the greatest amount of royal jelly is present. Grafting 1-day-old larvae into the queen cups produces the same result, with greater control over production. The frames are taken to royal jelly harvesting rooms. There, skilled labourers remove the larvae from the cups, discard them, and remove the royal jelly. The actual removal of the royal jelly from the cups is done by aspiration, mostly by a small vacuum pump that sucks the jelly from the cups into a reservoir (Figure 41-2). After it is collected it must be filtered, and because it spoils, it must be refrigerated (0–5°C), frozen (below –17°C), or freeze-dried under vacuum (lyophilized). Freeze-dried royal jelly powder can be stored for years.

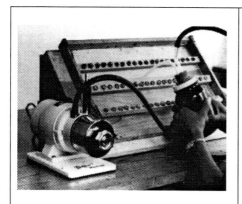

Figure 41-2. Collecting royal jelly from a special frame of queen cups. The bars have been rotated so the queen cups can be easily seen and the jelly harvested. A vacuum (pump, venture (water aspirator), or lung power) is applied to the tube through the two-hole stopper in the mouth of the reservoir. The vacuum sucks the royal jelly into the reservoir by the second tube that fits into the queen cups.

Table 41-1. The Chemical Composition of Royal Jelly with Comments on the Constituents (*dw* means of dry weight).

Chemical	Amount Present	Comments
Water	60 – 70%	
Proteins	17 – 45% dw	Proteins and glycoproteins.
Amino acids and peptides	about 2.5%	29 amino acids have been identified. All essential amino acids for honeybee and human nutrition are present.
Enzymes	small amounts	Phosphatase, cholinesterase, glucose oxidase, and others.
Sugars	18 – 52% dw	Fructose and glucose make up most. The composition of sugars is close to that expected in honey.
Lipids, including hydroxydecanoic acid (Chapter V – 16), cholesterol, phospholipids, and others	3.5 – 19% dw	Most of the lipids are as free fatty acids that have biologically unusual chemical structure. It is these that seem to give royal jelly its biological properties.
Minerals	2 – 3% dw	Potassium is prevalent, but other common minerals are present (Ca, Na, Zn, Fe, Cu, Mn, P)
Vitamins		Royal jelly is especially rich in vitamin B complex, with niacin most prevalent. Other vitamins (A, D, K) are absent, and C is only in trace amounts.
Inositol		Perhaps as part of phospholipid biochemistry.

Royal jelly has been collected for purposes such as cosmetics, lotions, dietary supplements, and research purposes in medicine. It is highly regarded by some people as an apitherapeutic material. Its benefits are reported to range from wound healing to improving complexion, countering emotional problems, stimulating metabolism, appetite, and weight gain in babies, and increasing sexual prowess. In general, those claims have not been adequately tested to be considered scientifically valid.

CHAPTER X – 42

PROPOLIS

Propolis is bees' glue and waterproofing material. The name is derived from the prefix "pro", which means before, and "polis" the city. That is because bees, mainly races of *Apis mellifera* from temperate climes, use it to close off the entrance to the hive. That provides protection against bad weather, cold, and invasion by enemies. Figure 42-1 shows a honeybee collecting gum from the leaf buds of a tree, and using the material in the upper entrance of a honeybee hive (Chapter VIII – 32) and as a general sealant.

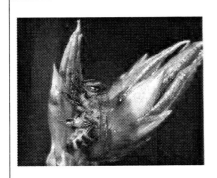

Figure 42-1. A honeybee gathering plant gum (**Left**) (note the corbicular load); it is made into propolis to reduce the entrance to the hive (**Centre**) to glue hive parts together (**Right**), and to generally waterproof and weatherproof the hive. It also has antimicrobial properties so may reduce the incidence of diseases in the colony.

Propolis ranges in colour from yellowy- to dark-brown, depending on plant origin. It remains sticky and pliable from 25°C to 45°C, but when it is cold, it becomes brittle. It often does not resume its soft texture on re-warming. At higher temperatures, propolis becomes increasingly sticky, and at about 60°C and higher it melts. Chemically it is highly complex: over 180 compounds are known from propolis in general. The most abundant materials are plant resins (flavonoids, phenolic acids and esters; about 55%) and gums or balsams (10%). Waxes make up about 30% of propolis, and much of that is from bees' wax. Essential oils may comprise 5–10%. Minor constituents, of which there are very many, originate from pollen (Chapter X – 40), or from the plants of origin.

It has long been known that propolis has medicinal value. Bactericidal and bacteriostatic activity is well proven. It has been found to inhibit the growth of various fungi. Antiviral effects are reported as well. It has been tested for cancer therapy, against ulcers, dental caries, bronchitis, and other ailments in people and other animals. It is reported to aid in the protection of wounds against infection, and has similar effects as tars when used to treat dermatitis.

Because of its value in apitherapy, there is a market for propolis. Naturally the bees use propolis to seal cracks and fissures, so it can be removed and collected during normal beekeeping and at honey harvest. It can be more purposefully collected from hives through the use of some sort of board or sheet ("trap") with narrow slots or holes (Figure 42-2). The bees naturally fill those crevices with propolis. To encourage the bees to fill the trap with propolis, the trap is deployed near the top of the hive, just inside the lid. The hive lid is propped open a crack to allow light and air to enter and so encourage the bees to seal the opening by filling the slots in the trap with propolis. The "traps" are removed from the hives. The propolis can be removed by scouring it from the trap with a special comb-like device, or, if the propolis trap is a flexible board with slot holes, it can be frozen and the brittle propolis removed by flexing, impact, or forced air.

Figure 42-2. A propolis trap that would be placed on the top of the hive, just under the lid.

Once raw propolis is collected, it must be processed. Propolis is soluble in alcohol, propylene glycol, and some vegetable oils. It can also be powdered. Thus, it has been incorporated into anti-chap lip balm, anti-inflammatory ointments, tinctures for topical application and for oral sprays against sore throat and bronchitis, shampoos, deodorant, soaps, and into tablets for ingestion.

Propolis has been used in making varnishes and polishes, and was thought to have been one of Stradivarius' secret ingredients for violin making. It has also been incorporated in plant grafting paste.

CHAPTER X – 43

BEE VENOM

Honeybees produce venom in the venom gland and store it in the poison sac at the base of their sting (see Figure 11-4). Venom secretion reaches the maximum when the adult workers are about 12 days old and ceases when they are about 20 days old. The reason that there is an interest in collecting venom is medical, for the treatment of bee sting allergy and in some other ailments, such as rheumatoid arthritis.

When a honeybee (*Apis mellifera*) stings, it injects about 0.15 to 0.3 mg of venom, the capacity of the venom sac. The median lethal dose (LD_{50}) for human beings is 2.8 mg of venom per kilogram of body weight. That is to say, the average 60 kg person has a 50% chance of surviving if stung by about 600 bees. However, most people who die, or are adversely affected by bee stings, suffer as a result of one to a few stings. That is because they are hyper-allergic. The symptoms are excessive swelling around the stung part of the body, spreading rapidly, accompanied by swelling of the body, especially around the mouth, neck and throat, then suffocation and heart failure. That sort of severe reaction is called anaphylaxis.

Hyper-allergenicity can arise unexpectedly, so beekeepers usually carry an Epipen® or similar device that injects a dose of adrenalin (epinephrine) to counteract the allergic reaction (Figure 43-1). People considering starting in beekeeping should be tested for hyper-allergenicity, but remember, the sting is meant to hurt and cause discomfort. Some swelling and pain are to be expected!

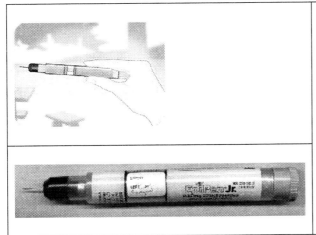

Figure 43-1. Epipen® kept by beekeepers in case of severe and adverse reaction to bee stings. The Epipen injects adrenalin into the body when the needle is jabbed into the thigh.

What is in venom? It is a clear, aromatic, acidic (pH 5.0–5.5) liquid with a sharp, bitter taste.   It dries quickly on exposure to air. It is used for pharmaceutical preparations in the treatment of various ailments and for desensitization of patients allergic to bee stings. Honeybee venom is more toxic than wasp venom. It is a complex mixture of chemicals, but mostly (88%) water. Table 43-1 present the composition and effects of the constituents. Although venom is mostly water, its other constituents are an array of chemicals that interact negatively with the integrity of the cells of the victim, and strong irritants, especially of the nerves, that can readily penetrate the damaged cells to bring about rapid discomfort. Many beekeepers become somewhat inured, or relatively non-reactive, to bee stings.

Table 43-1. The Constituents of Honeybee Venom
and Their Effects as Components of Stinging

Class of Molecule	Chemical	% in Dried Venom	Effect
Enzymes	Phospholipase - A	10 - 12	The enzymes attack the victim's cells, digesting cell membranes, exposing nerve cells, and causing vascular changes, itching, and pain.
	Hyaluronidase	1 - 3	
	Acid phosphomonoesterase	1	
	Lysophospholipase	1	
	Alpha-glucosidae	0.6	
Other proteins and peptides	Mellitin	50	Most of these are nerve poisons (neurotoxins) and irritants; they can also cause red blood cells to rupture.
	Apamine	1 - 3	
	Mast cell destroying peptide	1 - 2	
	Secapin	1	
	Procamine	1.5	
	Adolapin	1	
	Protease inhibitor	1	
	Tertiapin	tr	
	Small peptide	14	
Physiologically active amines	Histamine	0.5 - 2	These compounds cause vascular changes, itching and pain.
	Dopamine	0.1 - 1	
	Noradrenalin	0.1 – 0.7	
Amino acids	Aminobutyric acid	0.5	Possibly involved in nerve stimulation.
	Alpha-amino acids	1	
Sugars	Glucose & fructose	2	Probably neutral.
Phospholipids		5	
Other compounds		4 - 8	

In the past, venom was collected by surgical removal of the poison sac from worker honeybees. Nowadays, an electrified entrance board is placed into hives from which venom is to be collected (Figure 43-2). The apparatus consists of wires set about 6 mm apart and suspended 1 to 3 mm above a membrane of latex, plastic, silicon rubber, or taffeta (Figure 43-3). Every second wire carries an electrical charge of a maximum of 33 volts; the other wires are grounded. When the bees walk on the wires they are electrically shocked. That makes them sting through the membrane. Beneath the membrane is a glass plate (or absorbent material) that collects the venom. The dried venom can be scraped from the glass plate and placed into vials. Venom collected into absorbent material has to be washed out, and freeze-dried.

Figure 43-2. An old design of a venom collector. The honeybees are required to walk across an electrified entrance.

Figure 43-3. A venom collecting board, showing the wires, the membrane (above the glass plate, not shown), and many stings stuck into the membrane.

Colonies from which venom is collected become highly agitated by electro-shock treatment. Thus, venom collection should be infrequent and for a short duration. Some particularly defensive races of honeybees (e.g., Africanized bees, Chapter VI – 24) should not be used for venom collection. Venom production is a highly specialized undertaking, requiring extra protective clothing, including facemasks, fastidious cleanliness, and great care.

Bee sting apitherapy has been recommended to alleviate many human ailments; many of which have to do with inflammation of the joints and other tissues. It has its greatest reputation for relief of rheumatoid and other forms of arthritis and multiple sclerosis. Its use in treating hyper-allergenicity to bee stings is useful, especially for beekeepers who develop severe allergic reactions to being stung. By far the simplest way to receive bee sting apitherapy is to arrange to be stung.

CHAPTER X – 44

MORE BEES: QUEENS, COLONIES, & BROOD

Raising more bees is important to the beekeeping industry (Figure 44-1). The most important facet is raising queens. This is a skilled part of beekeeping that is taken on by specialists. They market their queens as individuals, in packages that contain 1 to 2 kg of worker bees, and in nucleus starter colonies. An ample supply of fertilized queens is needed for all.

Figure 44-1. Bees for sale! (**Left**) A **Benton queen cage** with queen and small retinue of a few workers (Chapter VIII – 31). (**Centre**) A **package of bees** showing screen walls and the syrup can being removed; the queen cage is suspended in the package by a tab beside the syrup can so that it can be grasped from the outside. (**Right**) A row of queen right **nucleus hives** (**nucs**) ready for shipment from Alberta.

Queen rearing was first commercialized by G. M. Doolittle (Chapter VIII – 31). His methods are still used today and are based on a thorough understanding of the early life history of brood (Chapter IV – 12 & 15).

The queen breeder selects hives with desirable characteristics as the **breeder hive**. There are several ways to raise queens, and one method is described briefly here. Within the selected hive, the queen is limited to one frame each day within the **confinement super**. Vertical queen excluders on each side of the frame are used to confine the queen, and the bottom of the chamber is solid. Usually the comb is dark so that the beekeeper can easily see the eggs laid in the bottoms of the cells. By restricting the queen to a single comb, the beekeeper is assured that all the eggs are the same age. The brood area is marked, and the eggs are allowed to hatch. That frame of brood is then removed from the breeder hive and the larvae, about 12 hours old, are ready for removal.

A **starter box** is prepared for receiving the larvae chosen to become queens. The starter box is supplied with frames of honey and pollen, and then with a large number of young nurse bees (Chapter IV – 15). The starter box is queenless and broodless. The entrance is blocked, but the bottom allows for ventilation (Chapters III – 10 & V – 19) so the bees are confined for several hours. Recognizing their queenlessness (Chapter V – 16), they become disposed to rear queens. At this time, the **grafts** are made (below).

The process of removing the tiny larvae from the breeder hive frame is called **grafting** (Figure 44-2). The larvae are removed on the end of a fine, hooked probe, the **grafting tool**, and placed into queen cups attached to the cell bar to be used in the queen production hive. The queen cups can be primed with a little royal jelly, thinned with water, so that the grafted larvae can be removed easily from the grafting tool. The frames with their bars (**Hoffman** or **graft frames**) of 12 queen cups are then placed into the populous and primed starter boxes. Three days later, those frames are removed to queenright **builder** or **finishing colonies**. Those colonies are equipped with a queen excluder across the bottom brood chamber (super), where the resident queen is confined. The grafted frame is placed with frames of unsealed worker brood, honey, and pollen, on either side in the super above. The young bees have more active hypopharyngeal glands and are more able to feed a large number of fast-growing larvae destined to be queens. They also have active wax glands for building and capping the queen cells as the larvae grow (Chapter IV – 15).

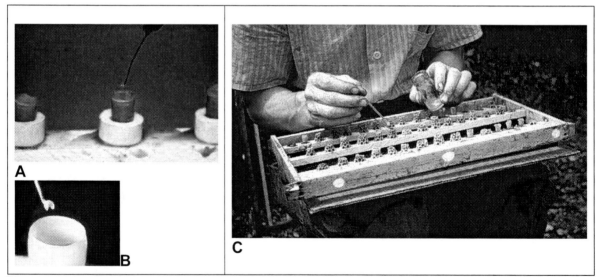

Figure 44-2. Grafting very young larvae into artificially made queen cups. (**A, B**) The grafting tool with a larva is gently inserted into the queen cup, primed with some diluted royal jelly. (The queen cups are on a small wooden base to allow them to be removed and reused.) (**C**) Some previously used (but soaked and washed) queen cups on a graft frame are being primed.

Thus, an abundant supply of nurse bees in strong and productive colonies must be maintained and encouraged to produce brood for queen rearing. A swarm box with 60 queen cells in it should produce about 50 queens. Overstocking the swarm boxes reduces production.

Ten days after grafting, the grafter frames are removed (Figure 44-3). The queen cells are removed individually and placed one per nucleus hive (usually fewer than five frames), now called a **mating nucleus hive**, or **mating nuc**. It is in those hives that the new, adult queens emerge.

Figure 44-3. (**Left**) Graft frames with capped queen cells ready for removal and (**right**) removed queen cells in vials ready to be taken to the mating nucleus hives.

The queens, in their mating nuclei, are taken to the **mating yard**, usually isolated from other colonies to allow for control in breeding (Chapter VI – 23). Some mating yards in larger operations may contain up to 2,000 nucleus hives. After the queens have taken their mating flights, they are ready for sale. They can be sold individually in Benton cages, in packages, or with the nucleus from which they have flown to mate. Queen breeders need to stock their mating yards with colonies that have desirable traits and that are producing lots of drones.

In summary, the sequence of events in the production of queens is:

Breeder Colony	⇒	Grafting	⇒	Starter Box	⇒	Finishing Colony	⇒	Mating Nuclei	⇒	Sale
4 days old				3 days		7 days		Emerge & Mate		$ $ $

Packages of bees are made by shaking 1, 1.5, or 2 kg of bees into a screen box. The process is simply a matter of shaking bees out of existing hives into a large funnel with the package below to receive the bees. Making packages is hot, stingy work! Usually, the boxes have four wooden sides and two screen sides. The queen and a small retinue of workers are placed in a Benton cage into a package box, and a feeder can of syrup forms a plug and provides food during shipment (see Figure 44-1).

Nucleus colonies, packages, and queen cages can be shipped by freight, mail, and some couriers. A cage of a kilo or so of buzzing bees can frighten some people. Queen producers can best advise how to ship their bees to their customers.

The package bee industry is mostly restricted to the southern USA in North America. Since the closure of the Canadian border to the USA because of mite diseases, Canadians are not able to import US bees. That has stimulated a domestic industry to supply mostly nucleus colonies to Canadian beekeepers, and has increased the importance of safely overwintering bees in Canada.

Bee brood can be sold as a commodity. In many non-European cultures, insects—including bee brood—is eaten as a source of protein in human diet. Where brood is reared specifically for use as a food, drone-size foundation is used for the combs, because the larvae and pupae in them are larger. The brood, when it is almost ready to pupate and is not yet capped, is simply knocked out of the frames onto a tray. Brood can be preserved by drying, freezing or canning. It can be eaten raw, fried or boiled. In some places, brood comb is eaten with brood in place. Bee brood is nutritious, containing about 15–18% protein, 3.5% fat, a little carbohydrate, and vitamins A and D. The levels of those nutrients are similar to those in beef or soybean.

International trade in bee brood is small, but I remember eating chocolate-coated baby bees. I am unable to find reference to the recipe now.

CHAPTER X – 45

POLLINATION

The **value** of pollination to agriculture has been estimated to be far greater than the value of hive products. Therefore, it is important for beekeepers and growers to understand pollination and the importance of bees. Providing pollination services to growers is hard work for beekeepers, but can be lucrative. Nonetheless, growers benefit more through additional crops of higher quality.

Pollination is simply the transfer of pollen from the anthers to the stigma of a flower (Figure 45-1). Self-pollination takes place if the transfer of pollen occurs within the same plant, sometimes within the same flower. In cross-pollination, two plants are involved. Self-sterile plants require cross-pollination (Figure 45-2), but other plants may be self-pollinating and self-fertile, and still others *require* no pollination. Pollination is the first step (after making a flower) in the sexual reproduction of plants. Pollination eventually gives rise to seeds, fruits, and the next generation of plants.

Fertilization is the step beyond pollination. **Pollen**, often produced copiously by the **anthers** of flowers (Figure 45-1), comprises individual grains. Each grain is, in fact, a microscopic plant which, as it germinates and grows, produces sperm cells. The female parts of a plant comprise the **stigma**. It is a special receptive surface for pollen. Below that in the flower is the style and then the ovary where the plant's egg cells (**ovules**) await fertilization (Figure 45-1). The pollen grains germinate on the receptive stigma and produce pollen tubes that grow through the stigma and down the neck-like **style** which ends at the ovary, enclosing the ovules (which, along with other cells, comprise a microscopic and protected embryonic plant within the parent plant).

Fertilization results from the merging of the sperm cells with the egg cells in the formation of an **embryo**, associated seed, and fruit which may contain from one seed (plum) to many seeds (pumpkin).

Pollen is rich in nutrients, proteins, carbohydrates, lipids, and vitamins (Chapters IV – 13 & X – 40). It is gathered by bees for food used in rearing brood. Other insects use it themselves as food for maturation. When insects gather pollen and nectar from flowers, plenty of pollen remains dusted on their bodies to be available for pollination.

Nectar is mainly a sugary liquid, but also contains small amounts of amino acids, minerals, lipids, and other constituents (Chapters IV – 13 & X – 38). It is produced by **nectaries**, mostly located at the bases of flowers. Nectar is the main source of carbohydrates for honeybees and other pollinators. Honeybees, bumblebees and a few other bees take nectar back to their nest (hives) where they cure it to produce honey (Chapter X – 38).

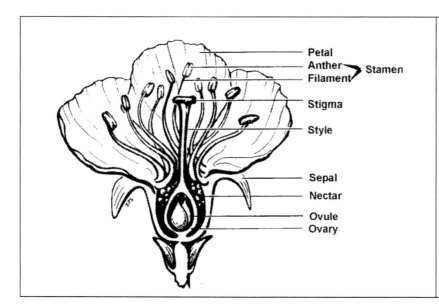

Figure 45-1. The parts of a flower. The nectar is secreted in droplets from nectaries. The pollen is shed as microscopic grains from the anthers.

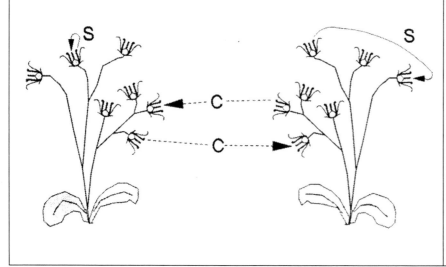

Figure 45-2. Cross (C) and Self-pollination (S). Self-pollination frequently occurs but does not result in fertilization unless the plant is a self-fertile species or variety. Many crop plants are self-compatible, but many are self-incompatible and require cross-pollination.

Insect pollinators are diverse. They include many kinds of flower-visiting insects that bring about the transfer of pollen. Bees are the most important pollinators to agriculture worldwide, but a large assemblage of flies, beetles, moths, butterflies, wasps and so on are involved in the process, especially in natural environments.

Many kinds of bumblebees (*Bombus*), and hundreds of kinds of solitary bees such as leafcutter bees, sweat bees, cuckoo bees, digger bees, carpenter bees, and even the specialized squash bee, are established in agricultural settings (e.g., the blueberry bee, Figure 45-5). All bees rely on flowers for pollen to feed their brood and on nectar for their own energy. Native bees were extremely important pollinators when fields were small and fencerows provided lots of nesting sites. In some instances they are still of major importance. However, modern agriculture now favours large fields, intensive mechanization, and use of chemicals for pest control (Chapter IX – 37), all of which are detrimental to natural pollinators. Although agricultural productivity is now far higher that it was, native bee populations have been reduced (Chapter XI – 51).

Domesticated bees have become the most important pollinators of crops. The European honeybee (*Apis mellifera*) is the most important, followed by the alfalfa leafcutter bee (*Megachile rotundata*) (also introduced from Europe). Bumblebees (*Bombus impatiens* is the one used most in North America) are now part of greenhouse tomato production. Other bees that show promise for crop pollination are orchard bees (*Osmia*), and perhaps the hoary squash bee (*Peponapis pruinosa*). All of the bees mentioned above, and others, can be manipulated artificially (SECTION XI).

Honeybees are now almost indispensable in the pollination of fruit and seed crops. Table 45-1 lists commonly grown temperate region crops and notes those that depend on, or benefit from, cross-fertilization by cross-pollination.

POLLINATION IN ORCHARDS. Cross-pollination is needed for many kinds of fruit trees because they are not self-fertile and require cross-pollination between varieties, not just between individual trees. The trees that provide a source of pollen compatible for the crop are called "**pollinizers**". Clearly the fruit from the pollinizer may be just as valuable and just as abundant as that from another cultivar that itself is probably a pollinizer as well. Orchard plantings need to take into account the pattern and mixture of cultivars grown. For example, in dwarf apple orchards, bees transport out-crossing pollen a distance of about 4 or 5 trees. The arrangement of varieties in orchards is crucial and should be considered when the orchard is first planted. The cultivars that donate and receive pollen must bloom in concert with each other to assure cross-pollination. Pollinizers can be provided by inter-plantings, top-grafting them, or by placing bouquets in the orchard during the blooming period. **Pollen inserts** are used extensively in some places. These are devices that attach to honeybee hives in such a way that the foragers must become dusted with pollen as they leave the hive. There are companies, mainly in the Northwestern USA, that specialize in harvesting and selling pollen from known cultivars of orchard trees. Certainly, without pollination some orchard trees (e.g., apples and pears) do not yield (Figure 45-3).

Timing of pollination is crucial. Fruit trees bloom at the end of spring. The blooming period is short but intense. Often the weather is cold, wet, and unfavourable for insect activity and pollination. Hence, wise growers reduce the risk of inadequate pollination by arranging for honeybees to be placed in the orchard.

Chapter 45
Pollination

The flowers of most fruit trees grown in Ontario are attractive to honeybees, and the general rule of thumb is to place strong and active hives in the orchard as early as possible when blooming starts. However, the flowers of pears (Figure 45-3) are not attractive to honeybees, so in this case some trickery is needed. For pollination of pears, it is best to introduce the bees when the orchard is about 25% in bloom. The bees, having been moved some distance by the beekeeper, forage close to their hives for the first few days and so visit pear blossoms and effect pollination. After a few days the bees become familiar with their surroundings and forage on distant and preferred flowers.

Placing and density of bees is important, e.g., how many colonies should be used and how they should be deployed. Colonies should be placed in sunny locations protected from cool winds, and preferably where the morning sun would warm them. (i.e., a south-facing place). Although recommendations often call for placing hives in small groups within orchards, that may not be practical. Recent research suggests that in orchards with flowers that are highly attractive to bees, honeybees forage avidly and widely so that spacing the hives may not be that important, except in pears (see above). As a general rule, adequate pollination is provided by two to three healthy hives of bees per hectare of producing fruit trees.

POLLINATION OF SMALL FRUITS requires similar management of honeybees. Raspberries, strawberries (to some extent), blueberries and cranberries all depend on, or at least benefit from, cross-pollination by insects (Table 45-1). Honeybees are excellent pollinators of raspberries and strawberries (Figure 45-6, left). Honeybees can pollinate the highbush and lowbush blueberries (Figure 45-5), but have difficulty with some varieties of highbush with very deep flowers. Honeybee colonies placed in blueberries often do not gain much weight because the flowers produce little nectar. Honeybees do not pollinate cranberries well (Figure 45-5). Cranberry flowers produce little nectar and their pollen is inaccessible to honey bees. Despite the value of bumblebees in cranberry pollination, honeybees are used as assurance against low populations of bumblebees and are placed on the fields (marshes) when the plants are about 10% to 25% in bloom (see pears).

Again, **placing and density of bees** is important. Recommendations for the numbers of colonies needed for pollination of small berries range from 2 to 10 hives/hectare (Table 45-1). The hives should be placed in the fields where the plants to be pollinated are growing. Spacing the colonies apart is probably not needed except where large areas are under production.

Elderberries and grapes do not benefit from the placement of honeybees where they are grown because honeybees generally do not visit these flowers, which provide only pollen and no nectar. Grapes (except wild species) are mostly self-pollinating.

Most **FORAGE LEGUMES** require insect pollination. Most bloom for at least a week or two during the summer. They produce nectar that is sought after by honeybees and is known to make excellent honey (Chapter X – 38). The blossoms of forage legumes are borne in inflorescences of many small flowers (sometimes called florets); each must be visited by a pollinator to set seed. The individual flowers are structurally complex, consequently some forage legumes are not well pollinated by honeybees (Figure 45-7). Sweet clover, alsike clover, and white clover are well pollinated by honeybees. Some cultivars of red clover have florets that are a little too long for honeybees to forage from: bumblebees are more adept (Figure 45-7, right). Honeybees pollinate bird's-foot trefoil, but as with red clover, they prefer other crops if they are available. Alfalfa is an important and highly valued source of honey for beekeepers, but honeybees are not generally effective at pollinating it. Honeybees are nectar thieves, removing nectar without affecting pollination. Alfalfa leafcutter bees (*Megachile rotundata*) are efficient at "tripping" and pollinating the flowers (Figure 45-7, left; Chapter XI – 46).

To bring about adequate pollination of the vast number of flowers (for example, half a billion flowers per hectare for red clover) of forage legumes, large populations of bees are needed. For honeybee pollination, three to seven colonies per hectare are recommended: for leafcutter bees, 50,000 cocoons per hectare of alfalfa.

POLLINATION OF OIL-SEED CROPS has become a serious business, especially for production of hybrid seed. Canola produces copious amounts of nectar which is eagerly sought after by honeybees. However, honey from canola readily crystallizes, often making it difficult to extract from the combs (Chapter X – 38). The pollination requirements for maximum production of canola crops are not understood. There are two species and many varieties of canola. The plants of the Argentine type are self-pollinating, and bee pollination is not regarded as a problem, but the plants of the Polish type require insects for cross-pollination and seed production. Much pollen probably moves between plants to bring about cross-pollination as the plants sway against each other in the wind. For hybrid seed production, two genetic lines of canola are grown in the same field, but in separate swaths. Most commonly, the genetic line from which seeds are harvested is male-sterile (lacks stamens) so it must be pollinated by a pollinizer line.

The old varieties of sunflower require insects to bring about pollination (Figure 45-6). Honeybees obtain large amounts of nectar from the blossoms. Newer, self-compatible varieties do not need insects for pollination, but they are still a good source of nectar. The honey is light amber and tasty. For soy beans, the generally held views are that cross-pollination is not needed and that the flowers produce no nectar. Both generalizations are wrong. At present, variety trials are needed to establish recommendations about the plants' pollination requirements and value for honey. Commercially grown flax does not need insect pollination for seed set.

POLLINATION REQUIREMENTS OF VEGETABLE CROPS are diverse. Several vegetables, such as pickling cucumbers (Figure 45-4, left), pumpkins and squash (Figure 45-4, right), and melons require pollination by insects. Honeybees are recommended at about 2.5 to 5 hives per hectare. Squash and pumpkin are well pollinated by native bees, notably the ground-nesting hoary squash bee (*Peponapis pruinosa*) in North America. Insects, especially bees, contribute to cross-pollination of eggplants and peppers; bumblebees are most effective (Chapter XI – 48). The flowers are buzz-pollinated by bees that vibrate (buzz) the flowers to remove the pollen which they collect (this also takes place in blueberries and cranberries): honeybees do not buzz-pollinate. Although tomatoes grown in the field set excellent crops without insect pollination, in the greenhouse that is not at all the case. Bumblebees (Chapter XI – 48) are now used worldwide for pollination of greenhouse tomatoes. Beans and peas are mostly self-pollinating. Pollination requirements for peanuts may differ between varieties, but self-pollination appears common. Vegetable seed production is a specialized area of agriculture that requires an understanding of pollination.

COLONY STRENGTH for pollination is important. Both beekeepers and growers need a yardstick by which to assess the value of colonies for pollination. Early in spring, colonies of honeybees have not yet become highly populous. So for orchard pollination, hives should have about 4,000 sq cm of live brood (that is, about 3 standard frames on both sides) with enough bees to cover 6 frames. Later in summer there should be more live brood, 5 or more frames on both sides, and enough bees to cover 10 frames in a colony for crop pollination. The colonies should have about 4.5 kg of honey or sugar syrup with them. The colonies should be free of American Foul Brood (Chapters IX – 33 & 34) and not be obviously infected with other diseases. Each colony should have a healthy, egg-laying queen. Colonies of those sizes may be in hives of a single full-depth super (single brood chamber), one and a half supers, or two supers. The number of supers, or boxes, in use does not affect the colony's effectiveness in pollination as long as there is room in the hive for the colony to expand and to prevent swarming. A colony of appropriate strength for pollination in only a single full-depth super has almost no room to expand. Beekeepers recognize the inconvenience and labour needed to move larger hives.

Moving Colonies has to be done at night, very early morning, or during very rainy weather. Growers and beekeepers need to coordinate their efforts to get the colonies to the crop at the best time and to remove them as soon as pollination is complete. As a general rule, beekeepers need about 36 hours' notice to move bees.

Pesticides, especially *insecticides*, are a hazard to honeybees (Chapter IX – 37). Beekeepers must be told about growers' intentions to apply insecticides.

Agreements for custom pollination are useful as they spell out the obligations and responsibilities of the growers and beekeepers with respect to delivery and removal of the colonies, their strength, their protection from hazards while on the crop, and the means of payment, availability of assistance, patterns of placement of colonies, and so on. Beekeepers and growers must enjoy mutual respect.

Table 45-1. Some Common Crop Plants Benefiting from Honeybee Pollination, Numbers of Colonies Recommended for Use, and Notes on Plants' Breeding Systems and Other Pollinators

Crop	Recommended Hive Density/ hectare	Comments
Apple	2.5	Self-incompatible within cultivars.
Apricot		Requirements not understood.
Blueberry (highbush)	5 or more	Some cultivars are self-incompatible, others not. Variability by cultivar. Pollination best accomplished by buzz-pollinating bumblebees and other wild bees.
(lowbush)	about 4	Self-incompatible pollination best accomplished by buzz-pollinating wild bees, including bumblebees.
Canola (Polish)	3 – 4	These old recommendations may or may not apply nowadays. Canola has undergone transformations in breeding and selection.
(Argentine)	-	
Cherry (tart)	1.5 – 2.5	Self-compatible, but insect pollination improves yield.
(sweet)	1.5 – 2.5	Self-incompatible.
Clover (red)	5 – 8	Bumblebees important, too.
(alsike, white and sweet)	5 – 8	Recommendations highly variable.
Cranberry	2.5	Pollination best accomplished by buzz-pollinating bumblebees.
Cucumber (field, pickling)	2.5 or more	Unisexual flowers, but plants self-compatible.
(salad long, greenhouse)	0	Pollination ruins the cucumbers.
Musk melon, cantaloupe	2.5	Unisexual flowers, but plants self-compatible.
Mustard	2.5 – 8	Self-incompatible.
Peaches	1	Recommendations conflicting, and plant pollination system incompletely known.
Pears	2 – 4	Self incompatible within cultivars. Special management needed (see chapter text).
Plums and Prunes	2.5	Requirements for cross-pollination variable between cultivars.
Pumpkins and Squash	1	Mostly served by *Peponapis pruinosa*.
Raspberries	1	Require insect pollination.
Strawberries	1 - 2	Some cultivars are self-compatible, some partially self-incompatible. Requirements need to be worked out variety by variety.
Sunflower	1 – 2	Even on self-compatible cultivars, cross-pollination improves yield.
Watermelon	1 – 2.5	Unisexual flowers, but plants self-compatible.

Some examples of floral mechanisms involved in pollination are shown in Figure 45-3 to 45-7.

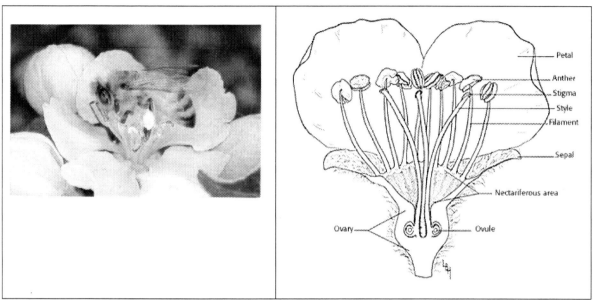

Figure 45-3. (**Left**) Pollination of apple: note how the honeybee is touching all the sexual parts of the flower. (**Right**) Cross-section of a pear blossom.

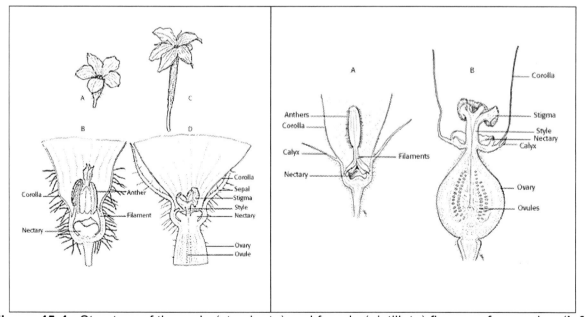

Figure 45-4. Structure of the male (staminate) and female (pistillate) flowers of cucumber (**left**) and squash or pumpkin (**right**). In these plants, both sexes of flowers grow on the same plant, but pollen must be transferred from staminate to pistillate flowers for pollination and fruit set.

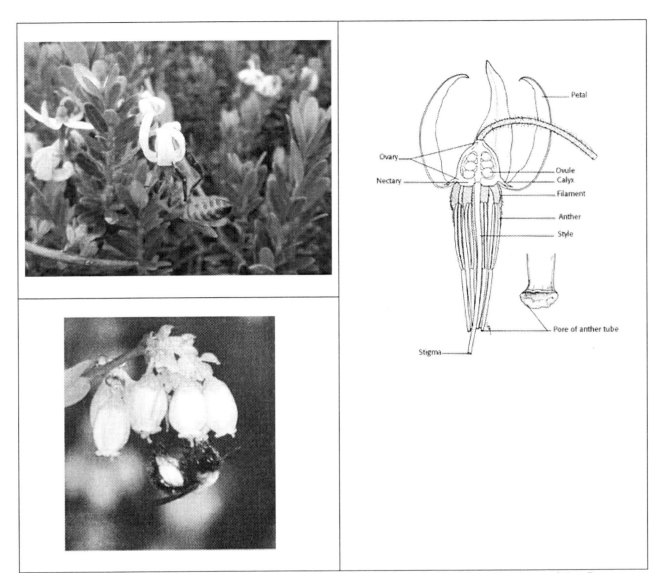

Figure 45-5. A honeybee visiting a flower of cranberry (**top left**). The structure of the flower requires pollinating bees to hang upside-down to obtain nectar and release pollen (see diagram on **right**). A wild blueberry bee, *Habropoda laboriosa*, pollinates a blueberry flower in the same general way (**bottom left**).

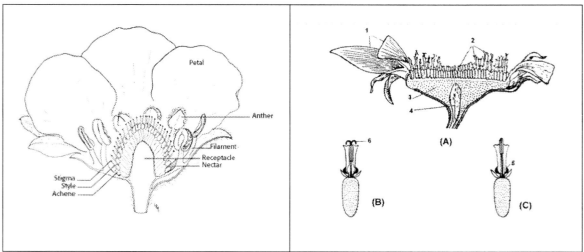

Figure 45-6. Flowers of strawberries (**left**) and sunflowers (**right**) are easily pollinated by honeybees crawling around the flowers, and among the mass of sunflower florets on a single head, while they gather nectar or pollen. On the sunflower, florets at different stages of development are shown (A2, B & C) along with the sterile ray flowers (A1).

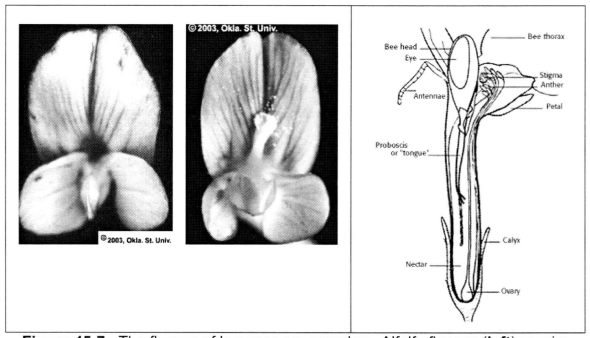

Figure 45-7. The flowers of legumes are complex. Alfalfa flowers (**left**) require "tripping" to release the pollen and expose the stigma. When leafcutter bees land on the keel, their weight pushes it down and it splits as the stigma and stamens are released. The release is so forceful that it deters honeybees from pollinating. On a red clover floret (**right**), the nectar is deeply hidden so bees with long tongues are adept at foraging and pollinating, e.g. Caucasian honeybees (Chapter VI – 24) and bumblebees.

Section XI

The Diversity of Managed Bees: Other Pollinators
Leafcutters, Alkali Bees, Bumblebees, Orchard Bees, Stingless Bees, & Conservation Needs

Although honeybees are the mainstay of pollination in agriculture, they do not service all crops well. This section briefly introduces the general life histories of alternative pollinators and honey producers.

Among the first alternative pollinators to be managed for pollination were the **alfalfa leafcutter bee** (*Megachile rotundata*) and the **alkali bee** (*Nomia melanderi*). The former is now the basis for an industry in "megachileculture" and serves not only for pollination of alfalfa (an important forage crop), but also for hybrid seed production for canola, and blueberry pollination.

Bumblebees are now reared in commercial operations, especially for pollination in greenhouse crops. Every time someone eats a greenhouse-grown tomato, a bumblebee is to be thanked. **Orchard**, or **mason**, **bees** (*Osmia* species) are used in various parts of the world for pollination of fruit trees. Several species are managed for that purpose. **Stingless bees** (Meliponini) are the indigenous **honeybees** of Central and South America. The Maya have descending gods, who bring heavenly gifts to humankind. One of them, called **Ah Muzencab**, carries stingless bee comb down from the heavens. The Meliponini are diverse, and many species have been encouraged to nest in boxes by meliponiculturalists who harvest their honey.

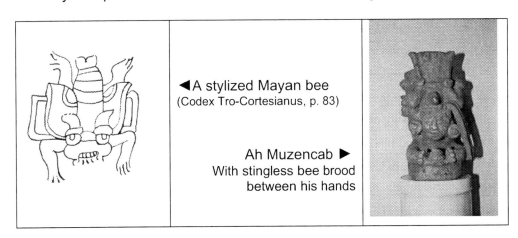

◀ A stylized Mayan bee
(Codex Tro-Cortesianus, p. 83)

Ah Muzencab ▶
With stingless bee brood
between his hands

Pollinator shortages have been known since ancient times, but recently scarcity and declines have become of grave concern for environmental sustainability, food security, and general conservation. The International Convention on Biological Diversity has recognized the problem, and pollinator conservation initiatives have begun globally. Pollination is crucial to agricultural and natural sustainable productivity and to the dynamic balance of the natural world.

CHAPTER XI – 46

ALFALFA LEAFCUTTER BEES

The alfalfa leafcutter (or leafcutting) bee, *Megachile rotundata*, is a hole-nesting bee. They are smaller than a honeybee and about 7 mm long. The females have their "pollen basket" or scopa of a dense brush of hairs underneath their abdomens, not on their legs as in most other bees (Figure 46-1).

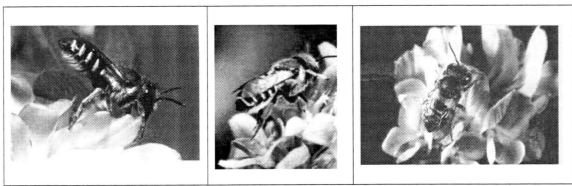

Figure 46-1. Alfalfa leafcutter bees on alfalfa flowers. Note the scopa of stiff hairs under the abdomen.

It appears that the alfalfa leafcutter bee entered North America from Eurasia in the 1940s and spread rapidly. W. P. Stephen was one of the scientists who first found nesting aggregations in holes in wood of an abandoned shed in Oregon and Idaho. It was there that he hand-drilled more holes in the sheds and placed out paper drinking straws set into plaster of Paris. The artificially created holes were taken up by the nesting bees. The pollinating activity of these bees on alfalfa had already been noted. Growers started their own encouragement of the bees with drilled wooden blocks and drinking straws in the 1960s. From those beginnings, the husbandry of alfalfa leafcutter bees has become a major industry, and critical to pollination of alfalfa. These bees are now being used for pollination of other crops, notably for hybrid canola seed production (Chapter X – 45), for other legumes, and for lowbush blueberries in the Maritimes.

Technological advances in keeping alfalfa leafcutter bees were made in Alberta, Canada, and the Northwestern USA. Developing the technology required a firm understanding of the biology of these bees. Male and female bees emerge from their overwintered cocoons in summer. They mate soon after, and the females seek out appropriate holes (e.g., beetle bore holes in wood) of between 6.4 and 7.2 mm diameter, and initiate their nest. They cut leaves (as their name suggests) using their large and powerful mandibles (as the name Megachilidea suggests from its Greek roots). They use those cut circles of leaves, and sometimes petals, to make a cup. That usually takes about 14–16 pieces of leaf. They can do that in about 2.5 hours.

The females then gather pollen, and collect some nectar from the flowers that they visit and pollinate (Figure 46-2 (A)). The pollen loaves they make are about 1/3 pollen and 2/3 nectar, and take them 15–27 trips to amass; about 5 hours of work on a good day. During that activity, the bees visit and pollinate thousands of flowers (sometimes called florets, Chapter X – 45). As she completes each cell, she lays an egg, and starts the next cell. The artificial tunnels used in alfalfa bee husbandry are 150 mm long and contain 10–12 cells when completed and plugged with more leaf discs (up to 50) pasted together. The female bee may provision cells in a second tunnel. On average she lives 40–50 days, but the males expire after about 15.

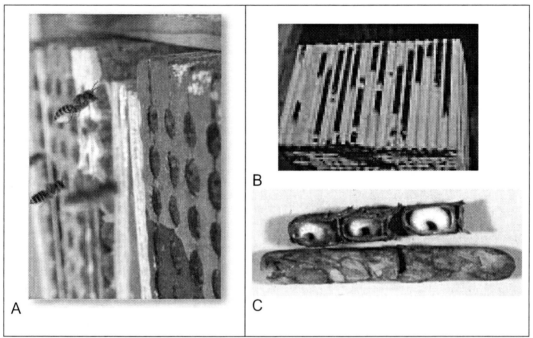

Figure 46-2. (**A**) Female alfalfa leafcutter bees coming home with scopa-loads of pollen. (**B**) The artificial nesting material, wood or polystyrene, is channeled so that two boards together make a set of tunnels. (**C**) The bees provision the leaf cells and lay eggs, and the larvae develop in a row. Males are produced near the entrance and emerge first. The females at the back emerge from the nest in the reverse order from which the eggs were laid.

Within the cells, the eggs hatch, and the larvae (Figure 46-2(C)) grow at a prodigious rate through four instars (Chapter IV – 12), and become pre-pupae. The duration for those events to unfold depends on temperature, but is between two weeks and a month. The pre-pupal stage is quiescent, and this is the stage during which the immature bees overwinter. When the cocoons are harvested by the alfalfa leafcutter beekeeper, the immatures are at this stage. The harvested cocoons are stored at cool temperatures (4°–6°C) that prevent further maturation. When they are needed the next year, temperatures of 29° to 31°C artificially substitute for natural incubation. Once incubated, the males take 20 or more days to eclose and emerge as adults, and the females have completed their development in a month. To provide pollination services, cells with pupae close to eclosion are placed in the field.

To make production reliable and smooth, storage and incubation are done in a single, temperature/humidity-controlled room, with some 30,000 to 40,000 cocoons placed on incubation trays (Figure 46-3, left). When the cocoons are ready to be shipped to the field, the cocoons may be shipped in the incubation trays or poured into pails (Figure 46-3, right).

Figure 46-3. (Left) Incubation tray with cocoons in temperature-controlled room. **(Right)** Pail of cocoons ready to be deployed for pollination.

In the field, alfalfa leafcutter bee shelters, with nesting boards, are in place when the cocoons arrive. There are many designs for shelters (Figure 46-4), and some are huge (road trailer–sized). The cocoons release their bees from the protection of the shelter. The males emerge first, then the females. They mate, and the cycle is complete.

Figure 46-4. Alfalfa leafcutter bee shelters: **(Right)** in Australia in recent trials there; **(Left)** on lowbush blueberry in New Brunswick.

CHAPTER XI – 47

ALKALI BEES

Alkali bees (*Nomia melanderi*) are ground-nesting Halictidae (Chapter VI – 25). They were observed as important pollinators of alfalfa in the 1940s, especially where the bees had formed natural aggregations of nesting females. The natural aggregations tend to last for a few years so were found not reliable for pollination services. W. P. Stephen and co-workers in Oregon had started to come to know their biology, and the types of soil they used for nesting. That was silt, perhaps with fine sand, less than 8% clay, neutral in acidity (the name of the bee not withstanding), and with high levels of salts ($NaCl$, $CaSO_4$, $CaCl_2$), and well drained. Their nesting requirements are special. With that knowledge, and more, Stephen and his co-workers designed and made artificial alkali bee beds (Figure 47-1). They "seeded" their beds with blocks of soil containing overwintering pre-pupae (Chapter IV – 12). Their artificial beds were colonized by dense aggregations of nesting bees, up to 2,000/sq m (about 50-fold more than the density in natural beds). The artificial beds stayed inhabited by nesting bees for at least 15 years. From large beds, the foraging bees spread over areas 4.5 km away, or 6,000 hectares, but not over continuous fields of alfalfa.

There are a few artificial bee beds still in use, but a variety of problems (pesticides, predators, pathogens, rainstorms, and economics of maintaining the beds) has caused steep decline in interest. Nevertheless, the story is interesting because it shows what can be done to address problems of pollinator shortages (Chapter XI – 51), and suggests that other ground nesters could be encouraged for pollination (Chapter XI – 50).

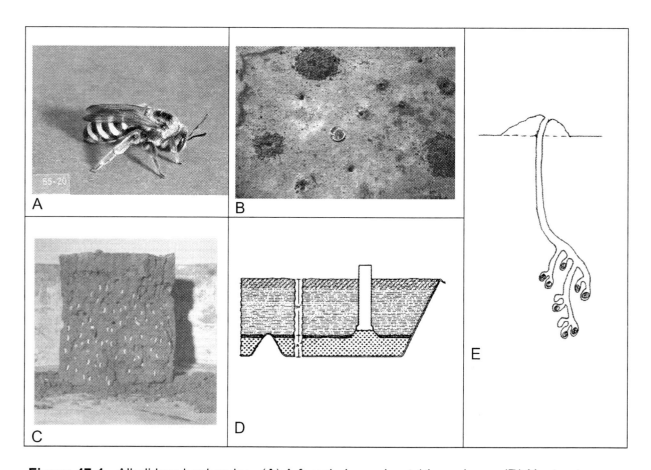

Figure 47-1. Alkali bee husbandry. (**A**) A female bee, about 14 mm long. (**B**) Nest entrances with fresh tumuli of excavated soil. (**C**) A block of soil with pre-pupae of alkali bees (white objects). (**D**) A plan for an artificial nesting bed. The whole bed is underlain with polyethylene sheeting above which is gravel and coarse sand. Most of the bed is the soil matrix over which the bees nest, but the top layer of 20 cm of saline soil is where they mostly nest. The downspout is for controlling water supply, and the mound in the base represents a latticework of soil that separates the whole bed into segments for individual management. (**E**) Drawing of typical nest, up to about 20 cm deep, with cells containing pollen loaves.

CHAPTER XI – 48

BUMBLEBEES

The value of bumblebees in pollination has long been recognized. Bumblebee husbandry for commercialized pollination services is rather new. Research into keeping bumblebees started in the late 1800s, especially through the work of F. W. L. Sladen, whose book *The Humble-Bee* remains a useful classic to this day. Bumblebees make annual colonies, started by a foundress, inseminated queen in the spring. The queen was inseminated the fall before. She emerges from her winter home, a hibernaculum under the ground, and starts searching for a nest site. Abandoned small rodent nests are usually chosen, but other places such as old upholstery in mattresses and furniture left outside are used. The queen organizes a cavity within the nest or stuffing, and makes a cup of wax that she secretes from her abdomen (Chapters II – 7 & X – 39), and lays in a stock of pollen. When she has enough, she lays several eggs (6–10), builds up and covers over the pollen and eggs with wax, and sets about incubating (Chapter V – 19) her brood at 30°–32°C, as would a bird. The eggs hatch in 4–6 days, and the larvae start to feed and grow prodigiously. Again, like many birds, the queen must go forth and forage for more pollen and nectar to feed and energize herself, and provide for her brood. The queen builds a jug-shaped honey pot in which to store nectar as an energy source and against bad-weather days when she cannot forage. The larvae take about 10–20 days to gain full size. They then pupate (Chapter IV – 12), spinning tough, silken cocoons. The pupal stage lasts about two weeks, then the first worker daughters eclose. Those workers assist the queen with expanding the nest, foraging for nectar and pollen, and rearing more brood from the eggs the queen lays. As the colony grows, the workers tend to be larger. The number of clutches of workers varies between species. Eventually, the queen produces eggs that are destined to become new queens and drones. The cues and mechanisms involved in the colonies' switching to production of sexuals is complex. At that time, the colony cohesiveness starts to deteriorate. The new queens and drones mate, the drones die, and the queens hibernate.

In selecting species of bumblebees for crop pollination, strong long-lived colonies are needed. The species must be amenable to captive breeding, and be naturally wide-ranging in the area where they are needed. The tricks of the trade involve the manipulation of the annual natural cycle so that bumblebees are under continuous culture, and that mating is successful. For the former, treatment of the inseminated queens with carbon dioxide simulates winter hibernation. It is important that they "buzz-" pollinate for use on some crops. The main crops for which bumblebee pollination is commercially successful are greenhouse tomato, egg plant, and peppers. Because of the expense and care required in producing large numbers of colonies of bumblebees, they have not been used extensively for outdoor crops. The species that have been commercially reared and successfully deployed in crop pollination are *Bombus impatiens* in eastern North America, *Bombus occidentalis* (until recently) in western North America, and *Bombus terrestris* in Europe and some other parts of the world. There is grave concern about risks posed by invasive species and about the trans-shipments of bumblebees around the world.

Figure 48-1. Bumblebees in pollination. **(A)** Hives deployed in a large greenhouse for tomato production in Ontario. **(B)** Hives amongst the tomato plants. **(C)** *Bombus impatiens* foraging on lousewort in the wild. **(D)** *Bombus terrestris* pollinating a tomato flower.

CHAPTER XI – 49

ORCHARD OR MASON BEES

Orchard bees are in the same insect family as alfalfa leafcutter bees, the Megachilidae. They are in a different genus, *Osmia*, and have some differences in their habits. They do not cut leaves, but use mud for nesting material to separate the cells. That habit has given them their other common name as mason bees. They nest in holes, as does the alfalfa leafcutter, and take readily to reeds, paper drinking straws, and cardboard tubes. In North America, the blue orchard bee (BOB) (*Osmia lignaria*) is slowly gaining popularity for pollination in orchards. It forages at temperatures (10°C) well below the threshold for honeybees, and has saved fruit crops that bloomed during cold weather when honeybees could not forage. In Japan, the horn-faced orchard bee (*O. cornifrons*) is used extensively for orchard pollination. In parts of southern Europe, notably Spain, the orange orchard bee (*O. cornuta*) has been tried, as has the red mason bee (*O. rufa*) in more northerly parts of Europe, notably the UK.

The life cycle of the orchard, or mason, bee is similar to that of the alfalfa leafcutter bee. However, their nest construction does not allow the cells and cocoons to be harvested *en mass*. BOBs produce only one generation each year. Their nesting and pollinating activities are naturally early in the year for about a month or six weeks, but individual females live only about 3 weeks. The tubes they like best are about 15 cm long and 7.5 mm in inside diameter. Artificial nests are used to manage these bees in orchards, and to keep them over the winter. The whole nest, or the tubes containing the bees, are kept at about 4°C for the winter, and then incubated at between 20° and 30°C. The higher the incubation temperature, the faster the bees emerge, but eclosion is usually complete within a week. The duration of the wintering period, region of origin of the bees (local populations are adapted to local conditions), and other factors influence the development rates of these bees. The aim of managing them is to assure they are available for pollinating the target crops: cherry, apple, almond, and others.

Artificial domiciles are made from blocks of wood with holes drilled into them, from cardboard tubes arranged inside milk cartons (they are waterproof), and cardboard tubes as sheaths to holes bored in wood. In Japan, bundles of reeds are used for the horn-faced orchard bee.

Figure 49-1. Orchard or mason bees. (**A**) The horn-faced orchard bee. (**B**) A blue orchard bee (BOB). (**C, D, E**) Three illustrations of orchard bee domiciles; the one in the tree is a bundle of cardboard tubes in a waxed box. (**F**) A fancy shelter with domiciles inside. Some birds peck at the domiciles to eat the larvae, so chicken wire screening is placed on a shelter to prevent that.

CHAPTER XI – 50

STINGLESS BEES

The stingless bees (Apidae: Meliponini) are such a diverse group (about 400 species are known worldwide) of eusocial insects that it is difficult to generalize about their biology and utility in human affairs. Certainly they are exploited extensively for their honey, especially by indigenous peoples of tropical and subtropical America, where they have their greatest diversity (300 species), and Africa. There are many ways of keeping stingless bees, but "bees in boxes" is a fitting generic description (Figure 50-1). Even so, a thorough understanding of their biology is needed to be a successful "meliponiculturalist".

Figure 50-1. A meliponiculturalist's house in Costa Rica. Note the bees in a box hanging from the tree, two gourd hives either side of the windows, a pot hive hanging over the verandah, a colony on the tree trunk, and an upright log hive to the right of the window on the right.

The stingless bees behave in some ways like solitary bees. They fully provision their cells with all the food the larvae need for development. That is called "**mass provisioning**". Once the queen (Figure 50-2) has laid an egg on the food mass (pollen, nectar, and some glandular secretions), the cell is closed and there is no further contact with members of the colony until eclosion. The **provisioning and oviposition process (POP)** is a complex of activities by which the queen finds an empty cell, and induces the workers to provision it.

While this POP is in progress, a worker may lay an egg into the cell. That egg, called a **trophic egg**, is eaten by the queen (presumably a protein meal supplemented with a nibble of provisions from the cell). Once the queen has laid her egg into the cell, the workers **fold over the collar of the cell** to tuck the egg away. There are many variations of the POP theme, depending on species. That is all rather different from what takes place in honeybees' lives, but the temporal sequence for division of labour is the same. The young workers first work with wax and cell-building material (cerumen), then some move to provisioning, and finally they become foragers.

Figure 50-2. A queen of a stingless bee. Note her hugely swollen abdomen, mostly occupied by her ovaries. She is an egg laying bio-machine.

The largest stingless bee is *Melipona flavipennis*. It is a little larger than a Western honeybee. The smallest are only a few mm long. The form of the bees varies greatly (Figure 50-3). The genus *Melipona* is robust, looking like small bumblebees or honeybees; many of the small "trigona" bees are slender and agile fliers.

Figure 50-3. Four common meliponine bees from Costa Rica. (**Top left**) *Melipona fasciata*; (**top right**) *Tetragonisca angustulata*; (**bottom left**) *Plebeia* sp.; (**bottom right**) *Partamona* sp. (Not drawn to scale).

Most species make their nests in existing cavities, such as hollow trees or underground. Some are associated with termite mounds, and a few construct their nest on open surfaces. The nests have a tunnel-like entrance. The shapes and lengths of the entrance, as well as the materials they use for the entrance, are highly variable (Figure 50-4). Some are simple decorations on the surface, others are long, horn-like tubes. Different species use different architecture.

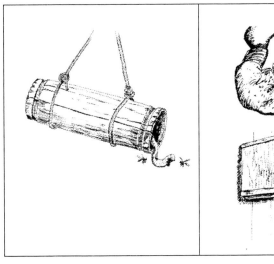
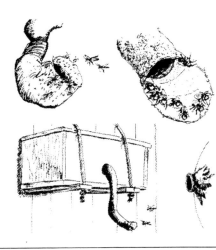

Figure 50-4. Some nest entrances. **Left** and **top left of 2nd panel** is *T. angustulata* (see Figure 50-3). **Top right** is of *Scaptotrigona pectoralis*. From the hive extends the entrance of *Nannotrigona mellaria*, and on the surface of the pot is the entrance of a *Plebeia* sp.

Within the nest (Figure 50-5), the bees use various building materials. Many species produce a hard batumen cover over the nest. The source of batumen is plant resin. Within the batumen cover, the bees build and maintain the involucrum. It is made of all sorts of materials, including their own wax, sand, mud, and chewed vegetable matter. Inside the involucrum is the brood chamber, with its horizontally arranged layers of cells. Also included within the nest, but separated from the brood chamber, are food storage areas. The honey pots can be huge in comparison with a cell of honeybee comb.

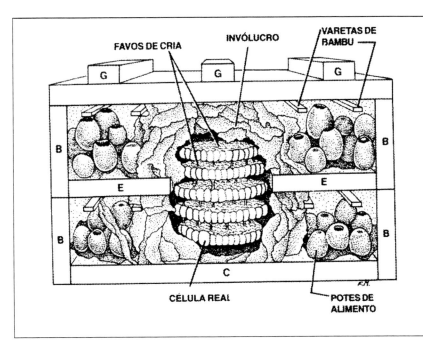

Figure 50-5. Inside a stingless bee's nest. There is no batumen depicted here, but the involucrum (*invólucro*) is clear. The brood area shows worker cells (*favos de cria*) and a queen cell (*célula real*). The food storage pots (*potes de alimento*).

CHAPTER XI – 51

OTHER POTENTIALLY USEFUL POLLINATING BEES & THE NEED FOR CONSERVATION

No one can deny the importance of honeybees in pollination, and in much of the world the Western honeybee is of prime importance. Nevertheless, honeybees are not always the answer to crop pollination. Other bees, such as alfalfa leafcutter bees, bumblebees, and orchard bees have proven their worth. The effort in research and development has paid off. What is needed is more effort placed into assessing other species of bees as potentially valuable. Already, the blueberry bee (*Habropoda laboriosa*) and the hoary squash bee (*Peponapis pruinosa*) have been identified as important contributors to crop pollination, but research into how to encourage them in agricultural settings has hardly started. Various researchers have identified dozens of other species of bees that need investigating. Even the stingless bees (Chapters VI – 25 & XI – 50) have only recently attracted attention as pollinators, mostly in Brazil and Mexico. Nesting boxes for carpenter bees (*Xylocopa* spp.) have been invented to encourage those bees to pollinate some crops in Malaysia and Brazil.

Recent applied ecological research in Canada, Mexico, USA, and Europe is now illustrating that communities of pollinators are more effective as a pollination force than are large populations of a single species. Diversity matters in ecosystem stability.

Meanwhile, pollinators and pollination have become recognized as being imperiled globally. The problems of pollinator shortages are well appreciated in the beekeeping industry. Various problems that have beset beekeepers (SECTION IX) have increased the difficulties, and the prices, for obtaining pollination services. That problem is not just limited to North America, but has resulted in less than maximum yields of various crops around the world. For example, apple production across the northern areas of the Indian subcontinent (e.g., in Pakistan, northern India, Nepal, and Bhutan) is below potential because of lack of pollination. Intensive agriculture is associated with sparse populations of pollinators in Canada, USA, and Europe. The reasons for those shortages are those raised for all declines in biological diversity: habitat destruction and fragmentation, poisons from pesticides, and increasing areas of monocultures. Ultimately, the problem comes down to human population pressures and rising consumer demands everywhere. With rising populations, the issue of food security is paramount. Yet agriculture depends to such a huge extent on pollination that the demise of pollinators must be seen as an exacerbating problem. It is reckoned that for human food, every third bite has required pollination and insect pollinators. The staples—rice, cereal grains, corn, and starchy root crops—can be regarded as independent of insects for pollination. The food animal chain on which carnivorous human beings depend has forage at its base. Forage requires pollination for seed production. In more natural settings, pollinators service the plants that produce the fruits and seeds eaten by wildlife such as migratory birds, and mammals from mice to hibernating black bears. Those same seeds assure the next generation of plants in natural succession.

Many problems confront conservation biology and sustainable ecosystem management, but pollination is central. The Brazilians have taken a lead role in pollinator conservation. Through their untiring efforts, the problem now has a much higher profile than it did just a decade ago. The Conservation on Biological Diversity (CBD or the Rio Convention) has recognized that pollination is an ecosystem service that can no longer be taken for granted. The North American Pollinator Protection Campaign is active in Canada, USA, and Mexico. The Europeans have made great strides in organizing and synthesizing the many examples of declines in pollinators that have occurred there. Reference to the European situation can be found at http://www.alarmproject.net/alarm/. There is an African Pollinators Initiative. Agencies around the world are calling for integrated scientific, economic, and political approaches to solving a problem that is just beginning to have major effects on people. Pollinator and pollination conservation initiatives have sprung up all over the world. They are striving to remedy some of the damage already done, and perhaps reverse the trends that are in place. The book "Status of Pollinators in North America" (2007) originated from the intensive deliberations of a highly qualified and international committee convened by the US National Research Council of the National Academy of Sciences lends credibility to the seriousness of pollinator conservation.

These are the last words in this book. Bees are fascinating, their husbandry is a source of wonder, their value to human beings through pollination and hive products is immense (billions of dollars worldwide), they are a crucial element in all but a few terrestrial ecosystems. Your personal apidological (and mellitological) knowledge is now far greater than that of the general public. Please keep in mind that for every third bite of your food, you should thank a bee!

FIGURE & TABLE CREDITS

Cover photographs from the author's collection and from Juli Paladino, Victoria MacPhail, and Sue Rigby.
Design by A. Morse.
Figure 1 - 1 available from many sources
Figures 1 - 2; 2 -1 to - 7; 3 - 2, 3, 5, 6, 8; 4 - 1 to - 3; 5 - 1; 6 - 1; 7 - 1, 2; 9 -1; 10 - 1; 11 - 2 to - 4, 6, 7; 12 - 1, 3, 4;
16 -1; 29 - 3; 40 - 1 redrawn and simplified from Snodgrass (1984) by A. Miciula
Figure 1 - 3 by A. Miciula
Figure 1 - 4 adapted from Sammataro and Propost (1998)
Figure 3 -1 Diagram based on Snodgrass (1984)
Figure 3 - 4 redrawn from Wootton (1990) and Dickinson (2001).
Figure 3 - 7; 18 - 5, 8; 20 -1, 4; 26 - 6; 26 -18; 31 - 17; 32 - 2, 3; 32 - 9 by P.G. Kevan
Cut-out Card used courtesy of the Smithsonian Institution Museum of Natural History, Washington DC.
Figures 7 - 3; 26 -1 simplified from Chapman (1998 and earlier editions) by P. G. Kevan
Figures 9 - 2, 3 diagrams by P. G. Kevan, simplified from various sources.
Figures 9 - 4, 5 diagrams by A. Miciula, simplified from various sources.
Figures 10 - 2; 11 - 1; 12 - 2; 13 - 1; 16 -2; 23 -1; 32 - 5, 6, 8; 33 - 1 to - 3; 34 - 2; 44 - 2, 3 courtesy of the University of Guelph
Figures 11 - 5; 20 - 5 (www.ars.usda.gov/Research/Research.htm?modeco...);
       33 - 4 www.ars.usda.gov/is/AR/archive/aug98/bees0898.htm (K8144-3)); and 34 - 6 courtesy of USDA
Table 12 - 1; 13 - 3; 15 - 1; 20 -1 by P. G. Kevan
Figures 13 - 2; 41 - 1, 2; 43 - 2, 3 courtesy of FAO and R. Krell (1996)
Figure 13 - 3 Photograph by Dennis L. Briggs courtesy of R. Thorp and www.vernalpools.org/Thorp/images/Thorp_23.jpg
Figure 14 - 1 adapted from Evans et al. (1984) by A. Miciula
Figure 14 - 2 simplified from Chapman (1998 and earlier editions) by Evans et al. (1984) and redrawn by P. G. Kevan
Figure 14 -3, 5 simplified from Evans et al. (1984) by P. G. Kevan and A. Miciula respectively
Figure 14 - 4 redrawn from Evans et al. (1984) by P. G. Kevan
Figure 15 -1 redrawn and combined from charts in Winston (1987) by A. Miciula and P. G. Kevan
Figure 15 - 2 redrawn from Winston (1987) by P. G. Kevan
Figure 16 - 1 redrawn and modified from Blum (1992) by A. Miciula
Table 16 -1; 35 - 1; 36 - 1; 37 - 1 information adapted from several sources and original literature by P. G. Kevan
Figure 16 - 3 redrawn and simplified from Winston and Slessor (1992) by A. Miciula
Chapter 17 portraits from www.nobelprize.org
Figures 17 -1 to - 3 redrawn from von Frisch (1967)
Table 17 -1 simplified from various sources by P. G. Kevan
Figure 17 - 4 adapted and redrawn from Gary (1992)
Figure 17 - 5 simplified from Lindauer (1957) and von Frisch (1967)
Table 17 - 2 summarized from von Frisch (1967)
Figure 18 - 1 redrawn from Heinrich (1976)
Figure 18 - 2 redrawn from Baldursson and Kevan (unpublished)
Figures 18 - 3, 4 redrawn from Waddington (1983)
Figure 18 - 6 by P. G. Kevan
Figure 18 - 7 by I. Smith in Kevan (1990)
Figure 18 - 8 redrawn from Laverty (1980)
Figure 19 - 1 redrawn from Seeley (1985)
Figure 19 - 2 based on Wilson & Milum (1927)
Figures 19 - 3 redrawn from Owens (1971) by A. Miciula
Figure 20 - 2 public photograph courtesy of static.flickr.com/22/26436086_437e8ca879_m.jpg
Figures 20 - 3; 31 - 14; 32 - 1 redrawn from Skinner, A., J. Tam and R. Bannister. 2006. Ontario Beekeeping Manual with an
      Emphasis on Integrated Pest Management.
Figures 21 - 1; 31 - 16 redrawn from Winston (1987) by A. Miciula
Figures 22 - 2, 3 redrawn by A. Miciula from data provide by Woyke (1996)
Figure 24 - 1 redrawn from Garnery et al. (1992) by A. Miciula
Figure 24 - 2; 31 - 15 by A. Miciula
Table 24 - 1 simplified from Ruttner (1987)
Figure 24 - 3 simplified from Ruttner (1987)
Figure 24 - 4 by A. Miciula
Figures 24 - 5; 25 - 5 to - 7adapted from information in Crane (1990) and other sources. Drawn by P. G. Kevan
Figure 24 - 6 adapted from various sources and drawn by A. Miciula
Tables 24 - 2; 34 - 1 summarized from various sources by P. G. Kevan

Figure & Table Credits

Figures 25 - 1 to - 3 simplified from Michener (2000) and other sources by P. G. Kevan

Figure 25 - 4 Courtesy of http://pikul.lib.ku.ac.th/insect/

Figure 25 - 8 redrawn from Koeniger et al. (1986)

Table 25 - 1 adapted from Seeley (1985) and prepared by A. Miciula and P. G. Kevan

Figure 26 - 2 adapted by A. Miciula from generally available diagrams and drawing

Figure 26 - 3 from web by C. Krebs www.macrophotos.com/charleskrebs/ck.html

Figures 26 - 4, 5; 27 - 1, 2; 29 - 2 simplified from Chapman (1998 and earlier editions), Barth (1985), and Wigglesworth (1965)

Figure 26 - 7 redrawn from Burkhardt (1964) and Chapman (1998)

Figures 26 - 8, 9 simplified from Backhaus (1993) by P. G. Kevan

Figure 26 -10 simplified from various sources by P. G. Kevan and A. Morse

Figure 26 - 11 from Kevan and Backhaus (1998)

Figure 26 - 12 by P. G. Kevan and M. Denny

Figure 26 - 13 redrawn from various sources, original research by von Frisch (1914).

Figure 26 - 14 redrawn and simplified from Gould (1985) and from Giurfa and Lehrer (2001)

Figure 26 - 15 redrawn from Lehrer et al. (1985)

Figure 26 - 16 adapted from various sources and from original photographs

Figure 26 -17 adapted from Daumer (1958)

Figure 27 - 3, 4 redrawn by A. Miciula from Barth (1985)

Figure 28 - 1 adapted by A. Miciula from generally available diagrams and drawing

Figure 28 - 2 by A. Miciula from various sources.

Figure 29 - 1 from Kevan and Lane (1985)

Tables 29 - 1; 31 - 1, 2 information from various sources compiled by P. G. Kevan

Figure 30 - 1 from Kevan and Lack (1985)

Figures 31 - 1 to - 3, 5, 6 from Crane (1983 and 2001)

Figure 31 – 4, 7, 8 to - 10 Copyrights expired and from various sources.

Figures 31 - 11, 12; 32 - 4 from Root, A. I. & E. R. Root (1908)

Figure 31 - 13 from Jenkyns (1886)

Figure 32 - 7 courtesy of Dadant & Sons Inc., 51 South 2nd, Hamilton, Illinois 62341, USA,
catalogue http://www.dadant.com/catalog/

Figures 32 - 10; 34 - 7 http://www.miteaway.com/Bee_Cozy_Wraps/bee_cozy_wraps.html and http://www.miteaway.com/
courtesy of NOD Apiary Products Ltd., P.O. Box 117, 2325 Frankford Road, Frankford, Ontario K0K 2C0

Figure 32 - 11 courtesy of the University of Guelph, and Root, A. I. & E. R. Root (1908)

Figure 34 -1 simplified from various sources by P. G. Kevan and courtesy of the University of Guelph

Table 34 - 1 simplified from various sources by P. G. Kevan

Figure 34 - 3 simplified from various sources and drawn by A. Miciula

Figure 34 - 4 simplified from various sources and drawn by A. Miciula, photograph from University of Guelph collection.

Figure 34 - 5 courtesy of M. T. Sanford and University of Florida Institute of Food & Agricultural Sciences.

Figure 35 - 1 Photograph on left courtesy of B. Patterson, right courtesy of Asociación Regional Apícola Canaria
C/La Solitica, 20, 38440. La Guancha. Tenerife, Islas Canarias. España
www.apicolacanaria.com/images/132_3227.JPG_0.jpg.

Figure 35 - 2 courtesy of the University of Bielefeld, Germany, www.uni-bielefeld.de/.../achroia-groß.jpg

Figures 35 - 3; 42 - 2 courtesy of www.doacs.state.fl.us/.../aethinatumida4.jpg (photography by J. Lotz) and
Dadant & Sons Inc., 51 South 2nd, Hamilton, Illinois 62341, USA, catalogue http://www.dadant.com/catalog

Tables 38 - 1; 39 -1; 41 - 1; 43 - 1 simplified from various sources, mostly from Krell (1996)

Figure 38 - 1 and Table 38 - 2 simplified from information in Crane (editor) (1975).

Figure 38 - 2 information available from many sources

Figure 38 - 3 redrawn from Kevan et al. (1992)

Figure 38 - 4 source unknown

Figure 39 - 1 courtesy of St. Brigid's, Kilburnie, Scotland, www.stbrigids-kilbirnie.com and Cheeky Bee
www.cheekybee.com/home.html

Figures 39 - 2, 3 courtesy of Colmenares Suizos, Chile www.colmenaressuizos.com

Figure 40 - 2 Courtesy of Dadant & Sons Inc., 51 South 2nd, Hamilton, Illinois 62341, USA, catalogue
http://www.dadant.com/catalog and University of Guelph

Figure 40 - 3 Courtesy of www.rawpower.com.au/beepollen.html

Figure 42 - 1 web, Krell (1996), and Pčelarstvo by Dr. Jovana Kulinčevića i Rajice Gačić III izdanje, BIGZ, Beograd,
www.pcelar.co.yu/propolis.htm

Figure 43 - 1 Courtesy of Epipen® www.epipen.com/images/epipen2.jpg.
Figure 44 -1 courtesy of University of Guelph and Chinook Honey Co., Box 12, Site 14, RR1

Figure & Table Credits

Okotoks, AB. T1S 1A1, Canada http://www.chinookhoney.com/gallery.htm
Figures 45 - 1, 2 from Kevan (1988)

Table 45 - 1 simplified and updated by P. G. Kevan from various sources
Figure 45 - 3 from web and courtesy of USDA McGregor (1976) http://gears.tucson.ars.ag.gov/book/chap5/pear.html
Figures 45 - 4 courtesy of USDA McGregory (1976) http://gears.tucson.ars.ag.gov/book/chap6/cucumber.html and
    http://gears.tucson.ars.ag.gov/book/chap6/pumpkin.html
Figure 45 - 5, top photograph courtesy of www.umaine.edu/.../HoneyBee-072704.JPG
    line drawing courtesy of USDA McGregor (1976) http://gears.tucson.ars.ag.gov/book/chap7/cranberry.html, and
    photograph lower left by J.A. Payne www.invasive.org/images/768x512/1224134.jpg
Figure 45 - 6 courtesy of USDA McGregory (1976) http://gears.tucson.ars.ag.gov/book/chap7/strawberry.html and
    http://gears.tucson.ars.ag.gov/book/chap9/sun.html
Figure 45 - 7 from North American Alfalfa Improvement Conference www.naaic.org/flower-06.jpg and courtesy of USDA
    McGregor (1976) http://gears.tucson.ars.ag.gov/book/chap3/red.html
Section XI illustrations from Codex Tro-Cortesianus www.famsi.org/mayawriting/codices/pdf/madrid.html and photograph by P. G.
    Kevan of replica from Museo de Antroplogía e Historía, Merida, Yucatan, Mexico
Figure 46 - 1 Courtesy of CSIRO www.ento.csiro.au/aicn/images/cain3117.jpg; and photographs by Karen Strickler on
    www.pollinatorparadise.com/Images/Bee_2a.jpg; www.pollinatorparadise.com/Images/Bee_1web.jpg.
Figure 46 - 2 photograph on left courtesy of G. Rank www.leafcuttingbees.com/images/leafcutter_bee.. ;photograph lower right
    courtesy of CSIRO www.ento.csiro.au/.../rr97-99/leafcutters.jpg
Figure 46 - 3 photograph on left courtesy of www.agr.gov.sk.ca/.../erualfalfa/SDAF46B.JPG
Figure 46 - 4 photograph on left courtesy of R. Bittner, photograph on the right, courtesy of the Government of
    New Brunswick www.gnb.ca/0171/10/images/b7a.jpg
    Figure 47 - 1 photograph on top left courtesy of Oregon State University
    www.ipmnet.org/kgphoto/images/55-20med.jpg others courtesy of W. P. Stephens (2003).
Figure 48 - 1 all photographs courtesy of BioBest Ltd. except lower right courtesy of John Haarstad Copyright © Cedar Creek
    Natural History Area (University of Minnesota), Cedar Creek. eebweb.arizona.edu/.../bumble/bombus_imp1.gif
Figure 49 - 1 Top left photograph by S. Batra courtesy of www.pollinatorparadise.com/Solitary_Bees. Bottom right courtesy of
    osmia.com/bluebee.htm
Figures 50 - 1, 3, 4 courtesy of K. Biesmeijer (1997)
Figures 50 - 2, 5 courtesy of P. Noguiera-Neto (1997)

References to printed literature from which material was drawn

Backhaus, W. 1993. Color vision and color choice behavior of the honey bee. Apidologie 24: 309 - 331.
Barth, F. G. 1985. Insects and Flowers: The Biology of a Partnership. Princeton University Press, Princeton, New Jersey, USA.
Biesmeijer, J. C. 1997. Abejas sin aguijón: Su biología y la organizacíon de la colmena. Elinkwijk BV, Utrecht, Netherlands.
Blum, M. 1992 Honey bee pheromones. In: Graham, J. M. (Editor). 1992. The Hive and the Honey Bee (Revised Edition). Dadant
    & Sons, Hamilton, Illinois, USA.
Burkhardt, D. 1964. Colour discrimination in insects. Advances in Insect Physiology 2: 131 - 174.
Chapman R. F. 1998. The Insects: Structure and Function. Pergamon Press, New York, USA (also material derived from earlier
    editions)
Crane, E. (Editor). 1975. Honey: A Comprehensive Survey. William Heinemann Ltd., London, UK.
Crane, E. 1983. The Archaeology of Beekeeping. Cornell University Press, Ithaca, New York, USA.
Crane, E. 1990. Bees and Beekeeping: Science, Practice and World Resources. Heinemann Newnes, Oxford, UK.
Crane, E. 2001. The Rock Art of Honey Hunters. International Bee Research Association, Cardiff, UK.
Daumer, K. 1958. Blumenfarben wie sie die Bienen sehen. Zeitschrift für vergleichende Physiologie 41: 49 - 110.
Dickinson, M. 2001. Solving the mystery of insect flight. Scientific American 284(6): 34-41.
Evans, H. E., J. W. Brewer, J. L. Capinera, R. G. Cates, G. C. Eickwort, & G. M. Happ. 1984. Insect Biology: A Textbook of
    Entomology. Addison Wesley Publishing, Reading, Massachussetts, USA.
Garnery L., J.-M. Cornuet, & M. Solignac. 1992 Evolutionary history of the honey bee Apis mellifera inferred from mitochondrial
    DNA analysis. Molecular Ecology,1: 145–154.
Giurfa M. & M. Lehrer. 2001. Honeybee vision and floral displays: from detection to close-up recognition. In: L. Chittka & J. D.
    Thomson (editors), Cognitive Ecology of Pollination. Cambrige University Press, Cambridge UK.
Gould J. L. 1985. How bees remember flower shapes. Science 227: 1429 – 1494.
Heinrich, B. 1976. Foraging specializations of individual bumblebees. Ecological Monographs 46: 105 - 128.
Jenkyns, C. F. G. 1886. A Book About Bees : Their History, Habits, and Instincts Together With the First Principles of Modern
    Beekeeping for Young Readers. Gardner, Danton, & Co., London, UK.
Kevan, P.G. 1988. Pollination, Crops and Bees. Ontario Ministry of Agriculture and Food. Publication 72. 13 pp.

Figure & Table Credits

Kevan, P.G. 1990. How large bees, *Bombus* and *Xylocopa* (Apoidea: Hymenoptera) forage on trees. Ethology, Ecology and Evolution 2: 233-242.

Kevan, P. G. & W. G. K. Backhaus. 1998. Color vision: ecology and evolution in making the best of the photic environment. Pp. 161 - 183. In: Color Vision: Perspectives from Different Disciplines (W. G. K. Backhaus, R. Kliegl, & J. S. Werner, editors). W. De Gruyter, Berlin, Germany.

Kevan, P.G. & A.J. Lack. 1985. Pollination in a cryptically dioecious plant, Decaspermum parviflorum (Lam.) A.J. Scott (Myrtaceae) by pollen collecting bees. Biological Journal of the Linnean Society 25: 319-330.

Kevan, P.G. & M.A. Lane. 1985. Flower petal microtexture is a tactile cue for bees. Proceedings of the National Academy of Sciences, U.S.A. 82: 4750-4752.

Kevan, P.G., H. Lee & R. Shuel. 1992. Sugar ratios in nectar of varieties of canola (*Brassica napus*). Journal of Apicultural Research 30: 99-102.

Koeniger N, Koeniger G, Gries M, et al. 1996. Reproductive isolation of Apis nuluensis Tingek, Koeniger and Koeniger, 1996 by species-specific mating time. Apidologie 27: 353 – 359.

Krell, R. 1996. Value-added Products from Beekeeping. Food & Agriculture Organization of the United Nations, Rome, Italy.

Laverty, T. M. 1980. The flower-visiting behaviour of bumble bees: Floral complexity and learning. Canadian Journal of Zoology 58: 1324 – 1335.

Lehrer M, R. Wehner, & M. V. Srinivasan. 1985. Visual scanning behaviour in honeybees. Journal of Comparative Physiology, A, 157: 405 – 415.

Lindauer, M. 1978. Communication among the Social Bees. Harvard University Press, Cambridge, Massachussetts, USA

McGregor, S. E. 1976. Insect Pollination of Cultivated Crop Plants. United States Department of Agriculture, Washington, DC, USA.

Michener, C. D. 2000. Bees of the World. Johns Hopkins Press, Baltimore, Maryland, USA.

. Noguiera-Neto, P. 1997. Vida e Criação de Abelhas Indígenas Sem Ferrão. Editora Nogueirapis, São Paulo, Brazil.

Owens, C. D. 1971. The thermology of wintering honey bee colonies. Technical Bulletin of the United States Department of Agriculture No. 1429, pp. 1 - 32.

Root, A. I. & E. R. Root. (1908). The ABC and XYZ of Bee Culture (1908 Edition). A. I. Root Co., Medina, Ohio, USA.

Ruttner, F. 1987. Biogeography and Taxonomy of Honeybees. Springer-Verlag, Berlin, Germany.

Sammataro, D. & J. Propost. 1998. The Beekeeper's Handbook. Cornell University Press, Ithaca, NY, USA

Seeley, T. 1985. Honey Bee Ecology. Princeton University Press, Princeton, New Jersey, USA.

Skinner, A., J. Tam and R. Bannister. 2006. Ontario Beekeeping Manual with an Emphasis on Integrated Pest Management. Ontario Beekeepers' Association, Bayfield, Ontario, Canada.

Snodgrass, R. E. 1984. Anatomy of the Honey Bee. Comstock Publishing Associates, Cornell University Press. Ithaca, NY, USA.

Stephens, W. P. 2003 In: Strickler, K. & J. H. Cane. 2003. For Nonnative Crops, Whence Pollinators of the Future. Thomas Say Publications in Entomology, Entomological Society of America, Lanham, Maryland, USA.

von Frisch, K. 1914. Der Farbensinn und Formensinn der Biene. Zoologische Jahrbuch Abt. allgemeine Physiologie der Tiere. 35: 1 - 182.

von Frisch, K. 1967. The Dance Language & Orientation of Bees. Harvard University Press, Cambridge, Massachussetts, USA

Waddington, K. 1983. Foraging behavior of pollinators. Pp. 213 - 239. In: Pollination Ecology (L. Real, editor). Academic Press, Orland, Florida, USA.

Wigglesworth, V. B. 1965. The Principles of Insect Physiology. Methuen and Co., London, UK.

Wilson, H. F. & V. G. Milum. 1927. Winter protection for the honey bee colony. Research Bulletin of the Wisconsin Agriculture Experiment Station No. 75: 1 - 47.

Winston, M. 1987. The Biology of the Honey Bee. Harvard University Press, Cambridge, Massachussetts, USA

Winston, M. & K. N. Slessor. 1992. The essence of royalty: Honey bee queen pheromones. American Scientist 80: 374 - 385.

Wootton,R.J. 1990. The mechanical design of insect wings. Scientific American 263(11):114-120.

Woyke, J. 1976. Polpulation genetic studies on sec alleles in the honeybee using the example of the Kangaroo Island bee sanctuary. Journal of Apicultural Research 15: 105 - 123.

Publications, mostly Books, suggested for additional reading and reference

Barth, F. G. 1985. Insects and Flowers: The Biology of a Partnership. Princeton University Press, Princeton, New Jersey, USA.

Bosch, J. & W. Kemp. 2001. How to Manage the Blue Orchard Bee as an Orchard Pollinator. National Agricultural Library, Beltsville, Maryland, USA

Buchmann, S. L. & G. P. Nabhan. 1996. The Forgotten Pollinators. Island Press, Washington, DC, USA

Canadian Association of Professional Apiculturalists. 1994. Honeybee Diseases and Pests. Canadian Association of Professional Apiculturalists, Calgary, Alberta, Canada.

Caron, D. M. 1999. Honey Bee Biology and Beekeeping. Wicwas Press, Cheshire, Connecticut, USA.

Chittka, L. & J. D. Thomson. 2001. Cognitve Ecology of Pollination. Cambridge University Press, Cambridge, UK.

Crane, E. (Editor). 1975. Honey: A Comprehensive Survey. William Heinemann Ltd., London, UK.

Crane, E. 1983. The Archaeology of Beekeeping. Cornell University Press, Ithaca, New York, USA.

Crane, E. 2001. The Rock Art of Honey Hunters. International Bee Research Association, Cardiff, UK.

Crane, E. 1990. Bees and Beekeeping: Science, Practice and World Resources. Heinemann Newnes, Oxford, UK.

Crompton, C. W. &. W. A. Wojtas. 1993. Pollen Grains of Canadian Honey Plants. Canada Department of Agriculture, Ottawa, Canada. Publication 1892/E. (similar publications are available for other parts of the world)

Delaplane, K. S. & D. F. Mayer. 2000. Crop Pollination by Bees. CABI Publishing, Wallingford, UK.

Esch, H. E. & J. E. Burns. 1996. Distance estimation by foraging honeybees. Journal of Experimental Biology 199: 155 - 162.

Frank, G. (editor). 2003. Alfalfa Seed & Leafcutter Bee Production and Marketing Manual. Irrigated Alfalfa Seed Producers Association, Brooks, Alberta, Canada

Free, J. B. 1982. Bees and Mankind. George Allen & Unwin, London, UK.

Free, J. B. 1993. Insect Pollination of Crops (2nd edition). Academic Press, London, UK.

von Frisch, K. 1967. The Dance Language & Orientation of Bees. Harvard University Press, Cambridge, Massachusetts, USA

von Frisch, K. 1971. Bees: Their Vision, Chemical Senses and Language. Cornell University Press, Ithaca, New York, USA.

Goodman, L. 2003. Form and Function in the Honey Bee. International Bee Research Association. Cardiff, UK.

Gould, J. L. & C. G. Gould. 1988. The Honey Bee. Scientific American Library, New York, USA

Graham, J. M. (Editor). 1992. The Hive and the Honey Bee (Revised Edition). Dadant & Sons, Hamilton, Illinois, USA.

Gruszka, J. 1998. Beekeeping in Western Canada. Alberta Agriculture, Food and Rural Development, Edmonton, Alberta, Canada.

Publications, mostly Books, suggested for additional reading and reference

Heinrich, B. 1981. Bumblebee Economics. Harvard University Press, Cambridge, Massachussetts, USA.

Hodges, D. 1984. Pollen Loads of the Honeybee. International Bee Research Association, Cardiff, UK. Johansen,

Johansen, C. A. & D.F. Mayer. 1990. Pollinator Protection: A Bee & Pesticide Handbook. Wicwas Press, Cheshire, Connecticut, USA.

Kevan, P.G. 1990. How large bees, *Bombus* and *Xylocopa* (Apoidea: Hymenoptera) forage on trees. Ethology, Ecology and Evolution 2: 233-242.

Kevan, P. G. (Editor). 1995. The Asiatic Hive Bee: Apiculture, Biology, and Role in Sustainable in Tropical and Subtropical Asia. Enviroquest Ltd., Cambridge, Ontario, Canada.

Krell, R. 1996. Value-added Products from Beekeeping. Food & Agriculture Organization of the United Nations, Rome, Italy.

Laverty, T. M. 1980. The flower-visiting behaviour of bumble bees: Floral complexity and learning. Canadian Journal of Zoology 58: 1324 – 1335.

Laverty, T. M. 1994. Bumble bee learning and flower morphology. Animal Behaviour 47:531 – 545.

Laidlaw, H. H. Jr. 1979. Contemporary Queen Rearing. Dadant & Sons, Hamilton, Illinois, USA.

Laidlaw, H. H. Jr. & R. E. Page. Jr. 1997. Queen Rearing and Bee Breeding. Wicwas Press, Cheshire, Connecticut, USA.

Lindauer, M. 1978. Communication among the Social Bees. Harvard University Press, Cambridge, Massachussetts, USA

McGregor, S. E. 1976. Insect Pollination of Cultivated Crop Plants. United States Department of Agriculture, Washington, DC, USA.

Michener, C. D. 1974. The Social Behavior of the Bees. Harvard University Press, Cambridge, Massachussetts, USA

Michener, C. D. 2000. Bees of the World. Johns Hopkins Press, Baltimore, Maryland, USA.

Moritz, R. F. A. & E. E. Southwick. 1992. Bees as Superorganisms: An Evolutionary Reality. Springer-Verlag. Berlin, Germany

Morse, R. A. 1983. A Year in the Bee Yard. Charles Scribner & Sons, New York, USA

Morse, R. A. (Editor). 1997. Honey Bee Pests, Predators and Diseases. A.I.Root, Medina, Ohio, USA.

Morse, R. & T. Hooper (editors). 1985. The Illustrated Encyclopedia of Beekeeping. E. P. Dutton, Inc., New York, USA

National Academy of Sciences of USA. 2007. Status of Pollinators in North America. The National Academies Press, Wsahington, DC, USA

Needham, G. R., R. E. Page, M. Delfinado-Baker, & C. E. Bowman. 1988. Africanized Honey Bees and Bee Mites. Ellis Horwood Ltd., Chichester, UK.
O'Toole, C. & A. Raw. 1991. Bees of the World. Facts on File, New York, USA.

Pellett, F. 1976. American Honey Plants. Dadant & Sons, Hamilton, Illinois, USA.

Root, A. I. 2007. The ABC and XYZ of Bee Culture: An Encyclopedia Pertaining to Scientific and Practical Culture of Bees (41th Edition). A. I. Root Co., Medina, Ohio, USA.

Roubik, D. 1989. Ecology and Natural History of Tropical Bees. Cambridge University Press, Cambridge, UK.

Ruttner, F. 1987. Biogeography and Taxonomy of Honeybees. Springer Verlag, Berlin, Germany

Seeley, T. 1985. Honey Bee Ecology. Princeton University Press, Princeton, New Jersey, USA.

Smith, D. R. (Editor). 1991. Diversity on the Genus Apis. Oxford & IBH, New Delhi, India

Snodgrass, R. E. 1984. Anatomy of the Honey Bee. Comstock Publishing Associates, Cornell University Press. Ithaca, New York, USA and London, UK.

Strickler, K. & J. H. Cane. 2003. For Nonnative Crops, Whence Pollinators of the Future. Thomas Say Publications in Entomology, Entomological Society of America, Lanham, Maryland, USA.

Towne, W. F. & W. H. Kirchner. 1989. Hearing in honey bees: Detection of air-particle oscillations. Science 244: 686-688

Townsend, G. F. & H. T. Heimstra. 2006. History of Beekeeping in Ontario. Ontario Beekeepers' Association, Bayfield, Ontario.

Vickery, V. R. 1991. The Honey Bee: A Guide for Beekeepers. Particle Press, Westmount, Canada.

Wenner, A. M. & P. H. Wells 1990. The Anatomy of a Controversy: The question of a" language" among bees Columbia University Press, New York, USA

Whiting, P. W. 1943. Multiple alleles in complementary sex determination of *Habrobracon*. Genetics 28: 365 – 382.

Wilson, B. 2004. The Hive: The Story of the Honeybee and Us. John Murray, London, UK.

Wilson, E. O. 1971. The Insect Societies. Harvard University Press, Cambridge, Massachussetts, USA

Winston, M. 1987. The Biology of the Honey Bee. Harvard University Press, Cambridge, Massachusetts, USA

Winston, M. 1992. The Killer Bee. Harvard University Press, Cambridge, Massachusetts, USA.

The main scientific journals for apidology

American Bee Journal

Apidologie

Bee Culture

Insectes Sociaux

Journal of Apicultural Research

There are also many magazines and newsletters devoted to beekeeping, and now there are also numerous sites on the world wide web from which information can be obtained on all aspects of bee biology and beekeeping.

Please be aware that information on the web is not necessarily reliable because there are no requirements for editorial or other review for posting.